MULTIHAZARD RISK ATLAS OF MALDIVES

Biodiversity—Volume IV

MARCH 2020

ADB

Notes:
In this publication, "$" refers to United States dollars.
The maps presented in this atlas reflect airports based on 2017 data from the Civil Aviation Authority of Maldives.

On the cover: An aerial view shows 1 of 26 natural atolls that make up Maldives, which also includes nearly 1,200 small coral islands and some of the world's most beautiful beaches. Recognized as the seventh-largest in the world, the coral reefs and associated ecosystems of Maldives are key foundations for food security and means of livelihood. Yet, they are considered as among the most vulnerable to climate change (photo by Roberta Gerpacio).

Contents

Tables and Maps

Tables

Maps

Foreword

Maldives is among the countries most vulnerable to the impacts of climate change as it is a small island nation with extremely low elevations. Maldives is also very vulnerable to impacts of rising air and sea surface temperatures and changes in rainfall patterns. Climate change impacts will therefore impose significant negative consequences on the Maldivian economy and society. Some of the priority vulnerabilities to climate change are land loss and beach erosion, infrastructure damage, degradation of coral reefs, and adverse impacts on water resources, food security, human health, and the overall economy.

Sustainable coastal resources management is of particular importance to Maldives, such that all regulations involving various development activities have coastal components. Despite the government's continued efforts in improving and sustaining coastal resources management, critical issues remain, such as the need for systematized coastal monitoring, clear definition of coastal boundaries and coastal development, enhanced regulatory and monitoring capacities for coastal resources protection, and sustainable long-term strategies on land reclamation and marine area protection. At a time when climate is rapidly changing and extreme weather events are frequently occurring, the critical roles that marine and coastal environments play in mitigating and adapting to climate change need to be sufficiently documented and properly recognized. It is therefore essential for Maldives to develop and establish a comprehensive digital database of marine and coastal ecosystem features and services that can be regularly monitored.

The *Multihazard Risk Atlas of Maldives* was developed through the project "Establishing a National Geospatial Database for Mainstreaming Climate Change Adaptation into Development Activities and Policies in Maldives" under the Asian Development Bank's regional knowledge and support (capacity development) technical assistance Action on Climate Change in South Asia (2013–2018). This five-volume atlas aims to promote the sustainable development of coastal and marine ecosystems and their various components, by enhancing the awareness of stakeholders on and enjoining them to address climate and disaster risks (including hazards, exposures, and vulnerabilities) to which ecosystems are exposed. The atlas presents spatial information and maps necessary for assessing future development investments in terms of their risks to climate and geophysical hazards.

The target audience of the *Multihazard Risk Atlas of Maldives* are the concerned stakeholders with current or planned development activities in the country, including public and private sectors, nongovernment organizations, research and academic community, development partner agencies, other financial institutions, and the general public. The atlas will also be a useful reference for other developing countries with similar geographical and environmental conditions, particularly small island developing states. It is envisioned that the atlas will significantly contribute to rendering important sector development investments more resilient to hazard-specific risk scenarios in the short, medium, and long terms.

H.E. Dr. Hussain Rasheed Hassan
Minister
Ministry of Environment, Malé

Shixin Chen
Vice-President for Operations 1
Asian Development Bank, Manila

Acknowledgments

Government Ministries, Departments, and Agencies in Maldives
Civil Aviation Authority
Land and Survey Authority
Marine Research Institute
Meteorological Service
Ministry of Economic Development
Ministry of Education
Ministry of Environment
Ministry of Fisheries, Marine Resources and Agriculture
Ministry of Health
Ministry of National Planning and Infrastructure
Ministry of Tourism
National Bureau of Statistics
National Disaster Management Center

International Institutions
Manila Observatory
Marine Spatial Ecology Lab, University of Queensland, Australia
SANDER + PARTNER
United Nations Development Programme

International Institutions in Maldives
International Union for Conservation of Nature, Maldives
United Nations Development Programme, Maldives

National Consultant Team
Ahmed Jameel, Integrated Coastal Zone Management Specialist
Faruhath Jameel, Geographic Information Systems Specialist and Team Leader
Hussain Naeem, Coastal Ecosystems and Biodiversity Specialist
Mahmood Riyaz, Climate Change Risk Assessment Specialist

Abbreviations

BODC	–	British Oceanographic Data Centre
CAA	–	Maldives Civil Aviation Authority
EPA	–	Maldives Environmental Protection Agency
GEBCO	–	General Bathymetric Chart of the Oceans
IHO	–	International Hydrographic Organization
IOC	–	Intergovernmental Oceanographic Commission
ME	–	Ministry of Environment
MED	–	Ministry of Economic Development
MFMRA	–	Ministry of Fisheries, Marine Resources and Agriculture
MLSA	–	Maldives Land and Survey Authority
MNPI	–	Ministry of National Planning and Infrastructure
MRC	–	Marine Research Centre
UNDP	–	United Nations Development Programme
UTM	–	Universal Transverse Mercator
WGS	–	World Geodetic System

Environmentally Sensitive Areas

Maldives is naturally blessed with an environment that supports a variety of terrestrial and aquatic life. As of 2017, it has a total of 284 environmentally sensitive areas. A quarter of these sites are for bird assembly, nesting, and habitat. Migratory birds find temporary refuge in the islands of Shaviyani Atoll. The mangrove ecosystem, which exists mostly in islands with the least human settlements, serves as breeding grounds for sharks and rays and as a nesting site for turtles. In addition to providing protection to the coast and coastal communities, mangroves also protect the white-breasted waterhen (*kabili*) and tortoise and serve as shelters for aquatic organisms.

Having evolved from corals, Maldives also has black, soft, hard, and table corals that shelter a variety of marine life. Divers visit Maldives to experience these beautiful coral formations and marine ecosystems teeming with life such as the yellowback fusilier, humpback snapper, whitetip shark, whale shark, grey shark, nurse shark, leopard shark, guitar shark, hammerhead shark, needlefish, eagle ray, manta ray, turtle, tortoise, sea cucumber, lionfish, moray eel, tuna, barracuda, sailfish, red snapper, grouper, and many others. Maldives also has local medicinal plants that the country aims to protect.

Rich biodiversity in Maldives. (Clockwise from top right) The *kabili*, also known as the white-breasted waterhen, is the national bird, while the grey heron is a common sight in some coasts (photos by Wang Chaonan and Anastasia Kolchedantseva). Under the sea, blacktip reef sharks and hawksbill sea turtles abound (photos by Ibrahim Rifath and Andrew Corman).

Map IV.1: Maldives, Environmentally Protected and Sensitive Areas

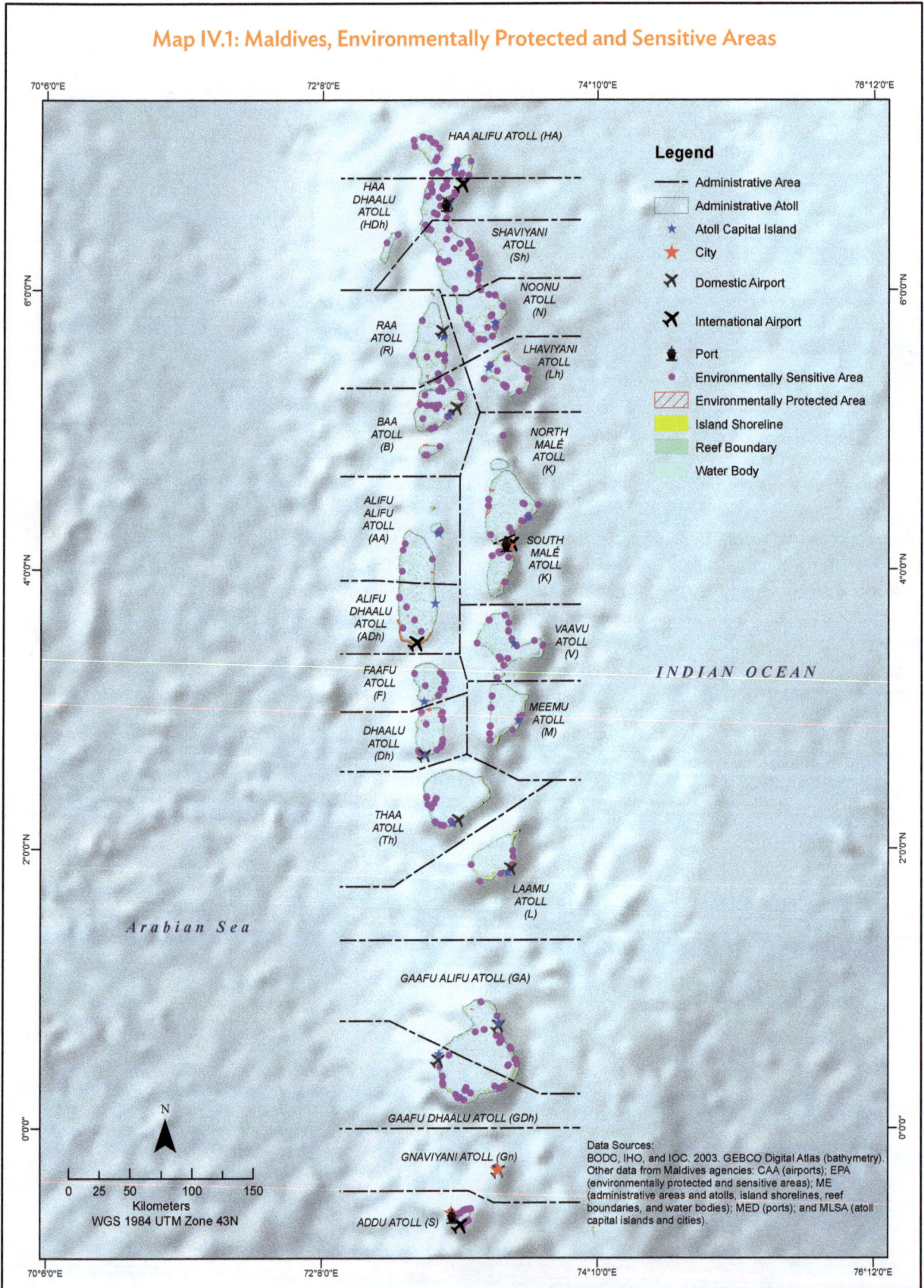

Legend
- Administrative Area
- Administrative Atoll
- ★ Atoll Capital Island
- ★ City
- ✈ Domestic Airport
- ✈ International Airport
- ⚓ Port
- ● Environmentally Sensitive Area
- ▨ Environmentally Protected Area
- Island Shoreline
- Reef Boundary
- Water Body

HAA ALIFU ATOLL (HA)
HAA DHAALU ATOLL (HDh)
SHAVIYANI ATOLL (Sh)
NOONU ATOLL (N)
RAA ATOLL (R)
LHAVIYANI ATOLL (Lh)
BAA ATOLL (B)
NORTH MALÉ ATOLL (K)
ALIFU ALIFU ATOLL (AA)
SOUTH MALÉ ATOLL (K)
ALIFU DHAALU ATOLL (ADh)
VAAVU ATOLL (V)
FAAFU ATOLL (F)
MEEMU ATOLL (M)
DHAALU ATOLL (Dh)
THAA ATOLL (Th)
LAAMU ATOLL (L)
GAAFU ALIFU ATOLL (GA)
GAAFU DHAALU ATOLL (GDh)
GNAVIYANI ATOLL (Gn)
ADDU ATOLL (S)

INDIAN OCEAN
Arabian Sea

N

0 25 50 100 150
Kilometers
WGS 1984 UTM Zone 43N

Data Sources:
BODC, IHO, and IOC. 2003. GEBCO Digital Atlas (bathymetry). Other data from Maldives agencies: CAA (airports); EPA (environmentally protected and sensitive areas); ME (administrative areas and atolls, island shorelines, reef boundaries, and water bodies); MED (ports); and MLSA (atoll capital islands and cities).

Map IV.2: Addu City, Environmentally Protected and Sensitive Areas

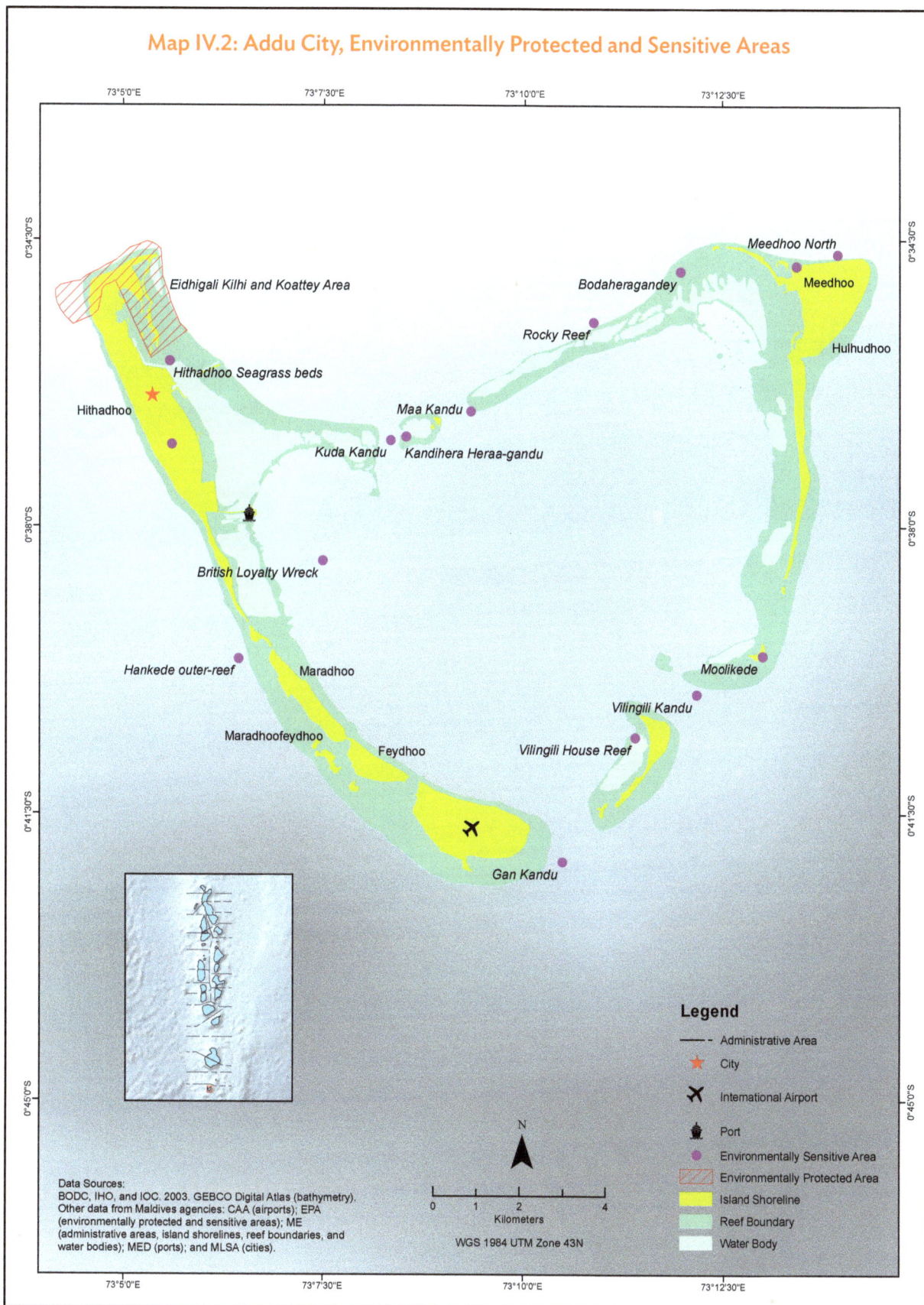

73°5'0"E   73°7'30"E   73°10'0"E   73°12'30"E

0°34'30"S

Meedhoo North

Eidhigali Kilhi and Koattey Area

Bodaheragandey

Meedhoo

Rocky Reef

Hulhudhoo

Hithadhoo Seagrass beds

Hithadhoo

Maa Kandu

0°38'0"S

Kuda Kandu   *Kandihera Heraa-gandu*

British Loyalty Wreck

Hankede outer-reef   *Maradhoo*   *Moolikede*

Vilingili Kandu

Maradhoofeydhoo   *Feydhoo*   *Vilingili House Reef*

0°41'30"S

Gan Kandu

0°45'0"S

Legend

— - Administrative Area

★ City

✈ International Airport

⚓ Port

● Environmentally Sensitive Area

▨ Environmentally Protected Area

▮ Island Shoreline

▮ Reef Boundary

▯ Water Body

N

0 1 2 4
Kilometers

WGS 1984 UTM Zone 43N

Data Sources:
BODC, IHO, and IOC. 2003. GEBCO Digital Atlas (bathymetry).
Other data from Maldives agencies: CAA (airports); EPA
(environmentally protected and sensitive areas); ME
(administrative areas, island shorelines, reef boundaries, and
water bodies); MED (ports); and MLSA (cities).

73°5'0"E   73°7'30"E   73°10'0"E   73°12'30"E

Map IV.3: Alifu Alifu Atoll, Environmentally Protected and Sensitive Areas

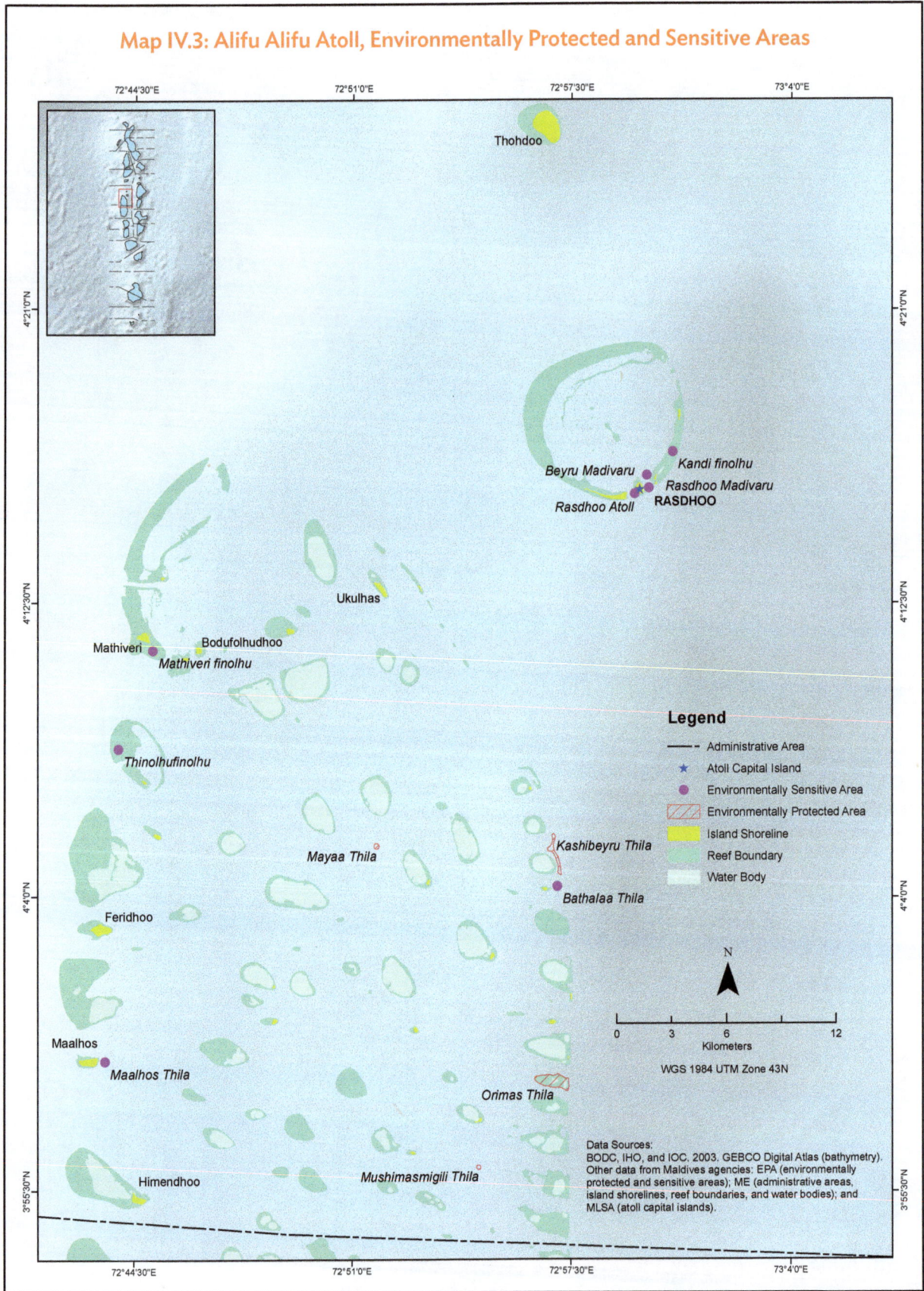

Thohdoo

Beyru Madivaru
Kandi finolhu
Rasdhoo Madivaru
RASDHOO
Rasdhoo Atoll

Ukulhas

Mathiveri
Bodufolhudhoo
Mathiveri finolhu

Thinolhufinolhu

Kashibeyru Thila
Mayaa Thila
Bathalaa Thila

Feridhoo

Maalhos
Maalhos Thila

Orimas Thila

Himendhoo
Mushimasmigili Thila

Legend

- – – Administrative Area
- ★ Atoll Capital Island
- ● Environmentally Sensitive Area
- ▧ Environmentally Protected Area
- ▮ Island Shoreline
- ▮ Reef Boundary
- ▮ Water Body

N

0   3   6   12
Kilometers

WGS 1984 UTM Zone 43N

Data Sources:
BODC, IHO, and IOC. 2003. GEBCO Digital Atlas (bathymetry).
Other data from Maldives agencies: EPA (environmentally
protected and sensitive areas); ME (administrative areas,
island shorelines, reef boundaries, and water bodies); and
MLSA (atoll capital islands).

Map IV.4: Alifu Dhaalu Atoll, Environmentally Protected and Sensitive Areas

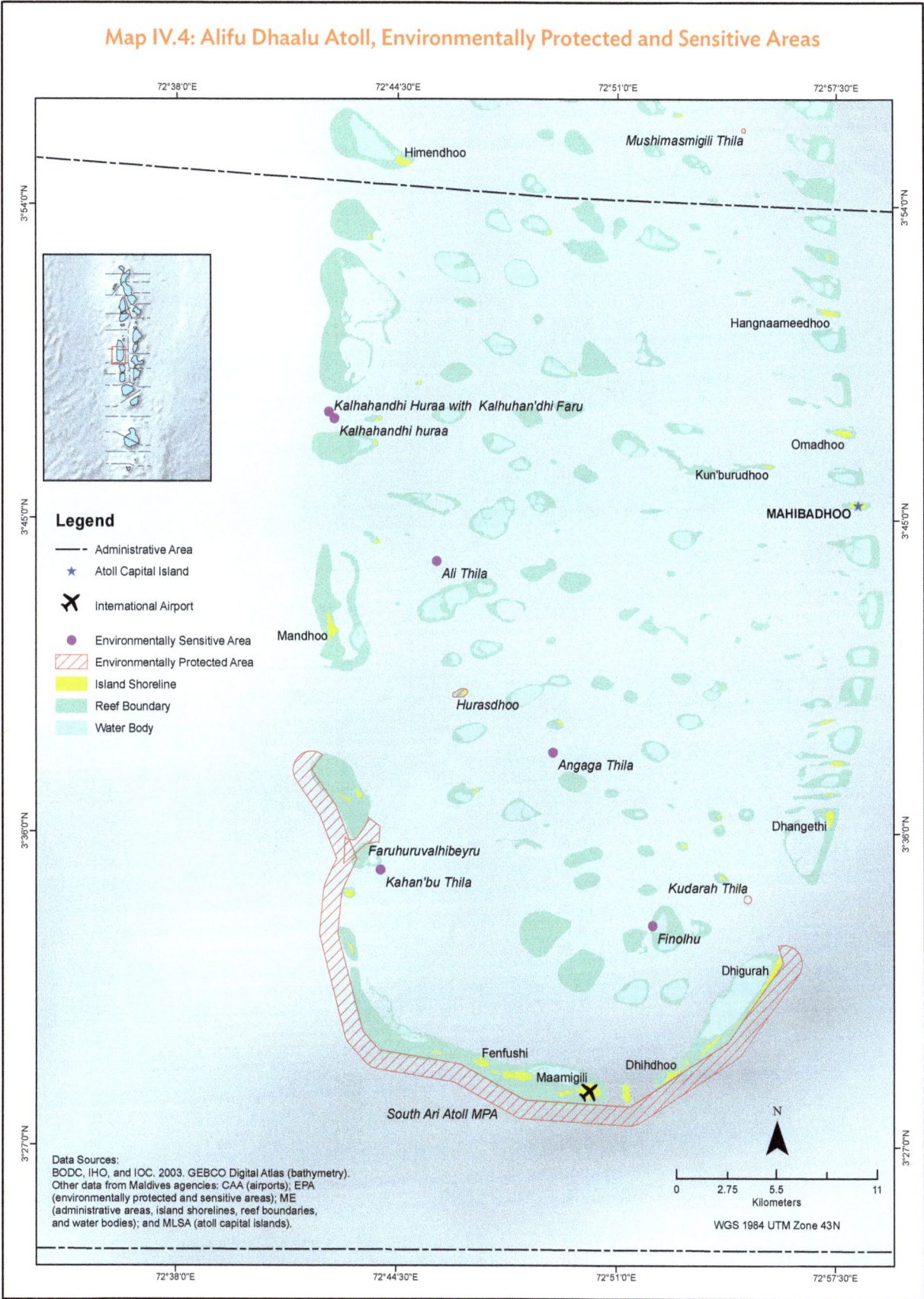

72°38'0"E   72°44'30"E   72°51'0"E   72°57'30"E

Mushimasmigili Thila

Himendhoo

Hangnaameedhoo

Kalhahandhi Huraa with Kalhuhan'dhi Faru
Kalhahandhi huraa

Omadhoo

Kun'burudhoo

MAHIBADHOO

Ali Thila

Mandhoo

Legend

- — Administrative Area
- ★ Atoll Capital Island
- ✈ International Airport
- ● Environmentally Sensitive Area
- ▨ Environmentally Protected Area
- ▮ Island Shoreline
- ▮ Reef Boundary
- ▮ Water Body

Hurasdhoo

Angaga Thila

Dhangethi

Faruhuruvalhibeyru
Kahan'bu Thila

Kudarah Thila

Finolhu

Dhigurah

Fenfushi   Dhihdhoo

Maamigili

South Ari Atoll MPA

N

3°54'0"N
3°45'0"N
3°36'0"N
3°27'0"N

Data Sources:
BODC, IHO, and IOC. 2003. GEBCO Digital Atlas (bathymetry).
Other data from Maldives agencies: CAA (airports); EPA
(environmentally protected and sensitive areas); ME
(administrative areas, island shorelines, reef boundaries,
and water bodies); and MLSA (atoll capital islands).

0   2.75   5.5   11
Kilometers

WGS 1984 UTM Zone 43N

Map IV.5: Baa Atoll, Environmentally Protected and Sensitive Areas

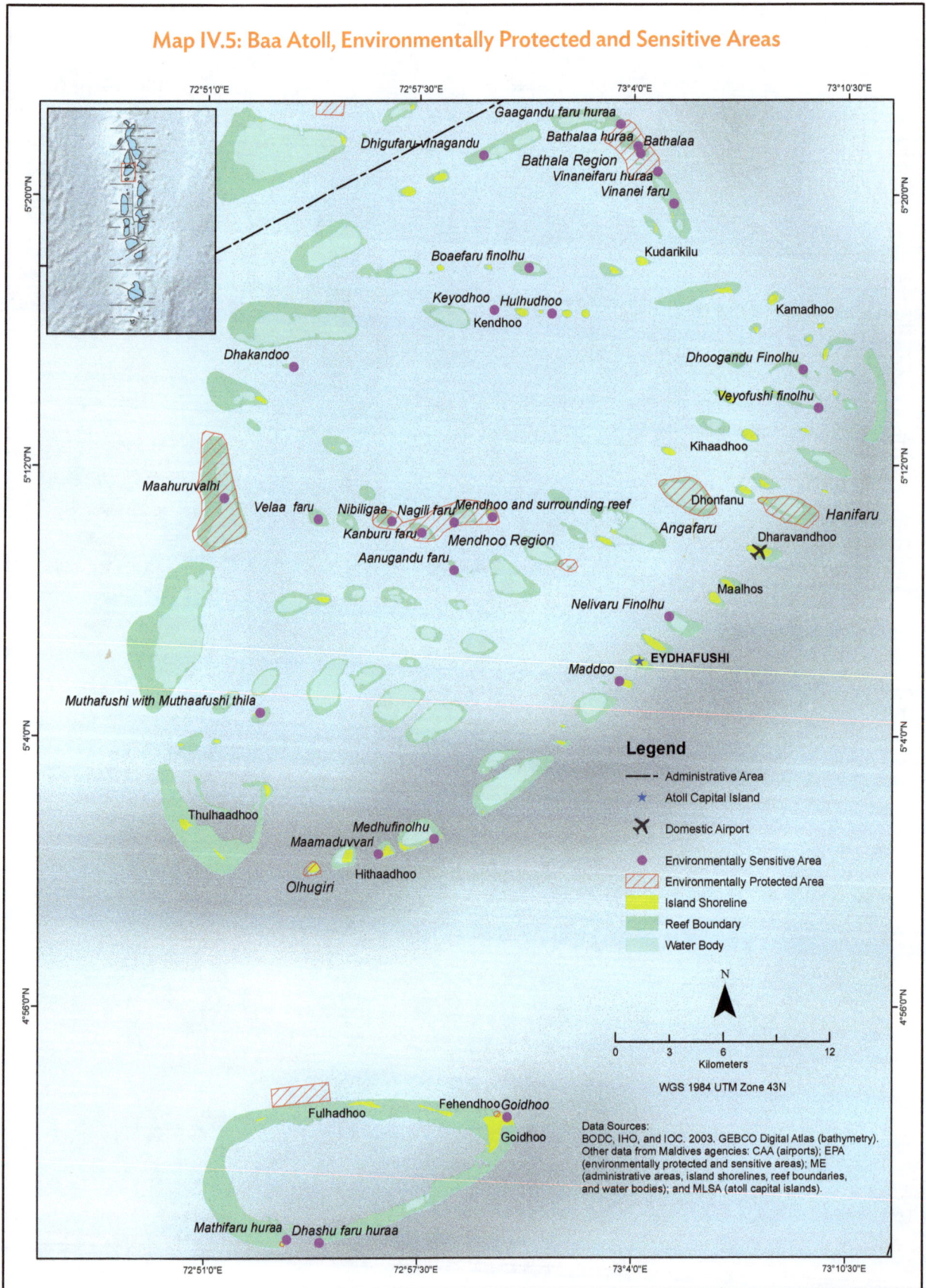

Gaagandu faru huraa
Dhigufaru vinagandu
Bathalaa huraa
Bathalaa
Bathala Region
Vinaneifaru huraa
Vinanei faru
Kudarikilu
Boaefaru finolhu
Keyodhoo
Hulhudhoo
Kendhoo
Kamadhoo
Dhakandoo
Dhoogandu Finolhu
Veyofushi finolhu
Kihaadhoo
Maahuruvalhi
Velaa faru
Nibiligaa
Nagili faru
Mendhoo and surrounding reef
Dhonfanu
Angafaru
Hanifaru
Kanburu faru
Mendhoo Region
Dharavandhoo
Aanugandu faru
Maalhos
Nelivaru Finolhu
Maddoo
EYDHAFUSHI
Muthafushi with Muthaafushi thila
Thulhaadhoo
Medhufinolhu
Maamaduvvari
Hithaadhoo
Olhugiri
Fehendhoo Goidhoo
Fulhadhoo
Goidhoo
Mathifaru huraa
Dhashu faru huraa

Legend

- – – Administrative Area
- ★ Atoll Capital Island
- ✕ Domestic Airport
- ● Environmentally Sensitive Area
- Environmentally Protected Area
- Island Shoreline
- Reef Boundary
- Water Body

N

0 3 6 12
Kilometers

WGS 1984 UTM Zone 43N

Data Sources:
BODC, IHO, and IOC. 2003. GEBCO Digital Atlas (bathymetry).
Other data from Maldives agencies: CAA (airports); EPA
(environmentally protected and sensitive areas); ME
(administrative areas, island shorelines, reef boundaries,
and water bodies); and MLSA (atoll capital islands).

Map IV.6: Dhaalu Atoll, Environmentally Protected and Sensitive Areas

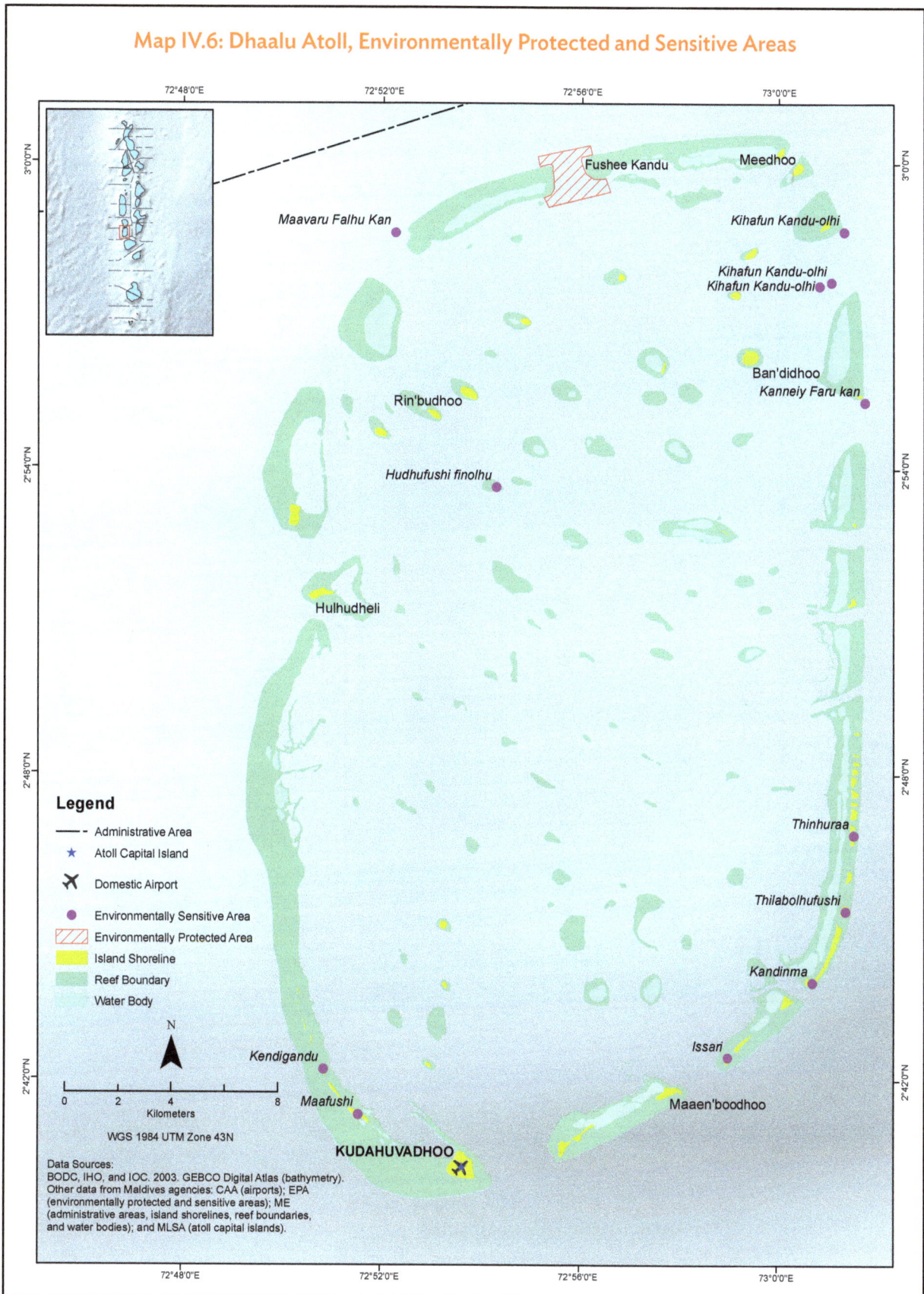

72°48'0"E 72°52'0"E 72°56'0"E 73°0'0"E

Fushee Kandu

Meedhoo

Maavaru Falhu Kan

Kihafun Kandu-olhi

Kihafun Kandu-olhi
Kihafun Kandu-olhi

Ban'didhoo
Kanneiy Faru kan

Rin'budhoo

Hudhufushi finolhu

Hulhudheli

Thinhuraa

Thilabolhufushi

Legend

— Administrative Area
★ Atoll Capital Island
✈ Domestic Airport
● Environmentally Sensitive Area
▨ Environmentally Protected Area
▪ Island Shoreline
▪ Reef Boundary
▪ Water Body

Kandinma

Issari

Kendigandu

Maafushi

Maaen'boodhoo

KUDAHUVADHOO ✈

N

0 2 4 8
Kilometers

WGS 1984 UTM Zone 43N

Data Sources:
BODC, IHO, and IOC. 2003. GEBCO Digital Atlas (bathymetry).
Other data from Maldives agencies: CAA (airports); EPA
(environmentally protected and sensitive areas); ME
(administrative areas, island shorelines, reef boundaries,
and water bodies); and MLSA (atoll capital islands).

Map IV.7: Faafu Atoll, Environmentally Protected and Sensitive Areas

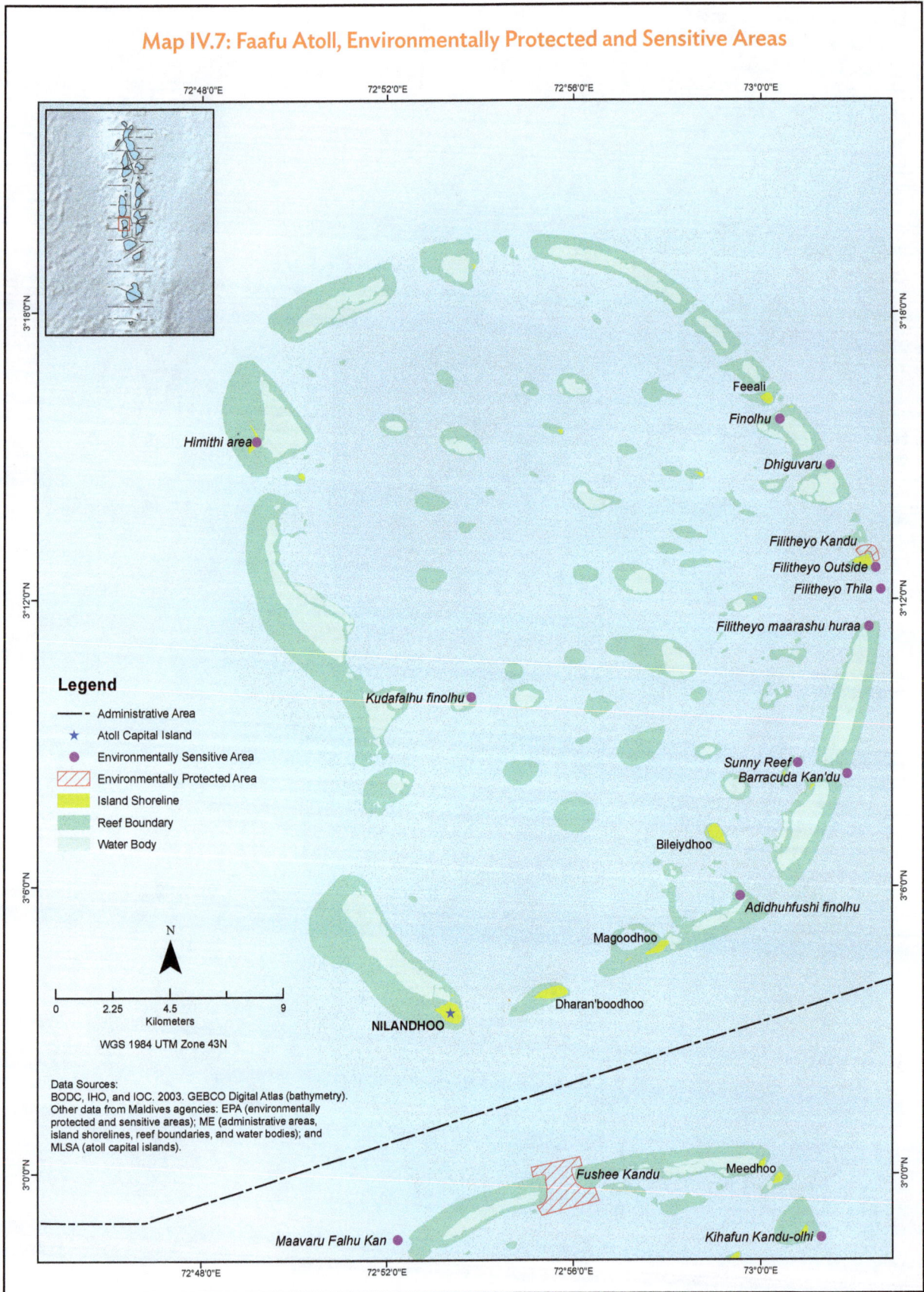

Feeali
Finolhu
Dhiguvaru
Himithi area
Filitheyo Kandu
Filitheyo Outside
Filitheyo Thila
Filitheyo maarashu huraa
Kudafalhu finolhu
Sunny Reef
Barracuda Kan'du
Bileiydhoo
Adidhuhfushi finolhu
Magoodhoo
Dharan'boodhoo
NILANDHOO
Fushee Kandu
Meedhoo
Maavaru Falhu Kan
Kihafun Kandu-olhi

Legend

- – – Administrative Area
- ★ Atoll Capital Island
- ● Environmentally Sensitive Area
- ▨ Environmentally Protected Area
- ▨ Island Shoreline
- ▨ Reef Boundary
- ▨ Water Body

N

```
0      2.25      4.5          9
        Kilometers
```

WGS 1984 UTM Zone 43N

Data Sources:
BODC, IHO, and IOC. 2003. GEBCO Digital Atlas (bathymetry).
Other data from Maldives agencies: EPA (environmentally
protected and sensitive areas); ME (administrative areas,
island shorelines, reef boundaries, and water bodies); and
MLSA (atoll capital islands).

Map IV.8: Gaafu Alifu Atoll, Environmentally Protected and Sensitive Areas

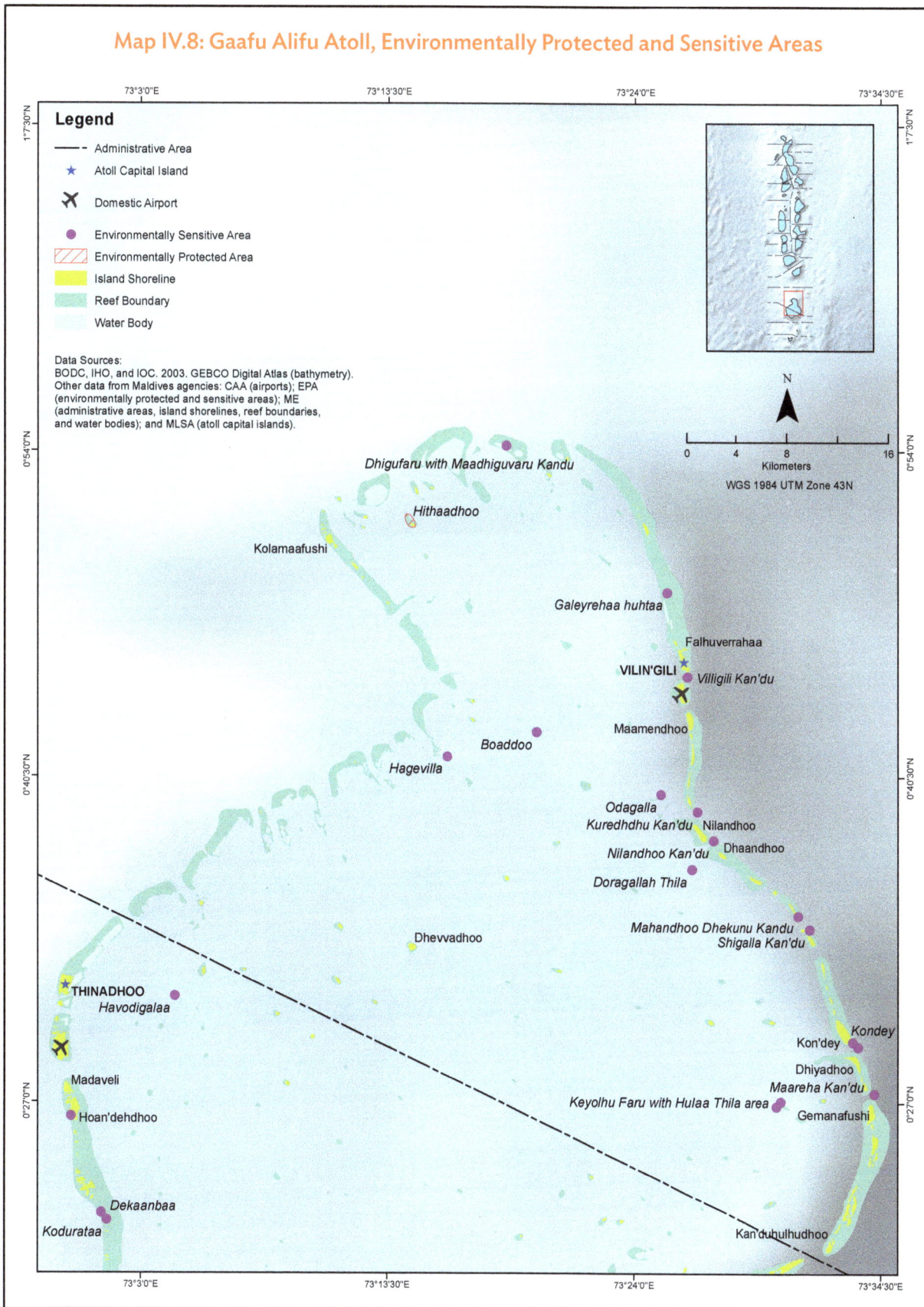

Map IV.9: Gaafu Dhaalu Atoll, Environmentally Protected and Sensitive Areas

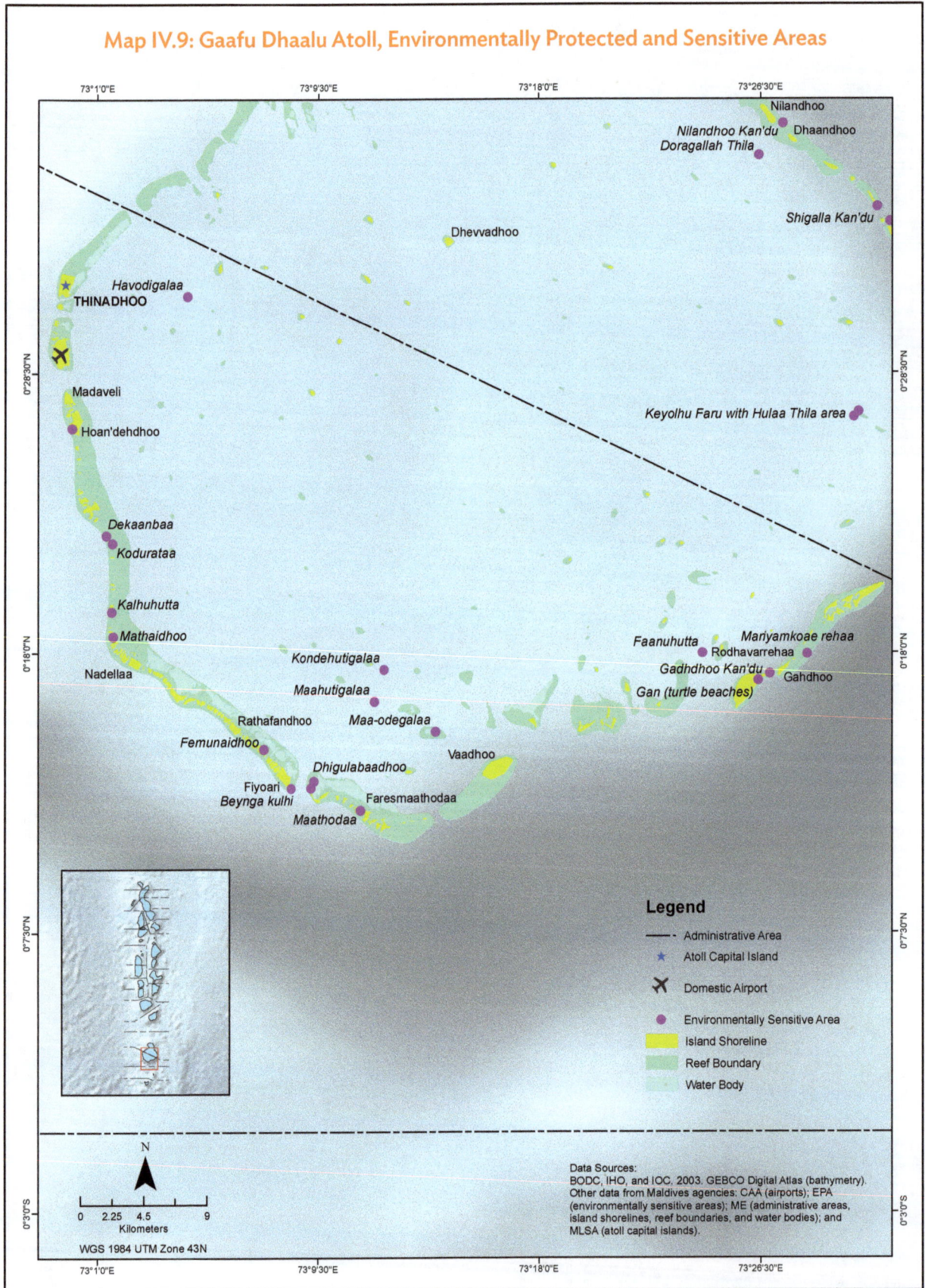

Nilandhoo
Nilandhoo Kan'du Dhaandhoo
Doragallah Thila

Shigalla Kan'du

Dhevvadhoo

Havodigalaa
THINADHOO

Keyolhu Faru with Hulaa Thila area

Madaveli
Hoan'dehdhoo

Dekaanbaa
Kodurataa

Kalhuhutta
Mathaidhoo
Nadellaa

Faanuhutta *Mariyamkoae rehaa*
Rodhavarrehaa
Gadhdhoo Kan'du Gahdhoo
Gan (turtle beaches)

Kondehutigalaa
Maahutigalaa
Rathafandhoo *Maa-odegalaa*
Femunaidhoo
Vaadhoo
Dhigulabaadhoo
Fiyoari
Beynga kulhi Faresmaathodaa
Maathodaa

Legend
— Administrative Area
★ Atoll Capital Island
✈ Domestic Airport
● Environmentally Sensitive Area
▬ Island Shoreline
▬ Reef Boundary
▬ Water Body

0 2.25 4.5 9
Kilometers
WGS 1984 UTM Zone 43N

Data Sources:
BODC, IHO, and IOC. 2003. GEBCO Digital Atlas (bathymetry).
Other data from Maldives agencies: CAA (airports); EPA
(environmentally sensitive areas); ME (administrative areas,
island shorelines, reef boundaries, and water bodies); and
MLSA (atoll capital islands).

Map IV.10: Gnaviyani Atoll, Environmentally Protected and Sensitive Areas

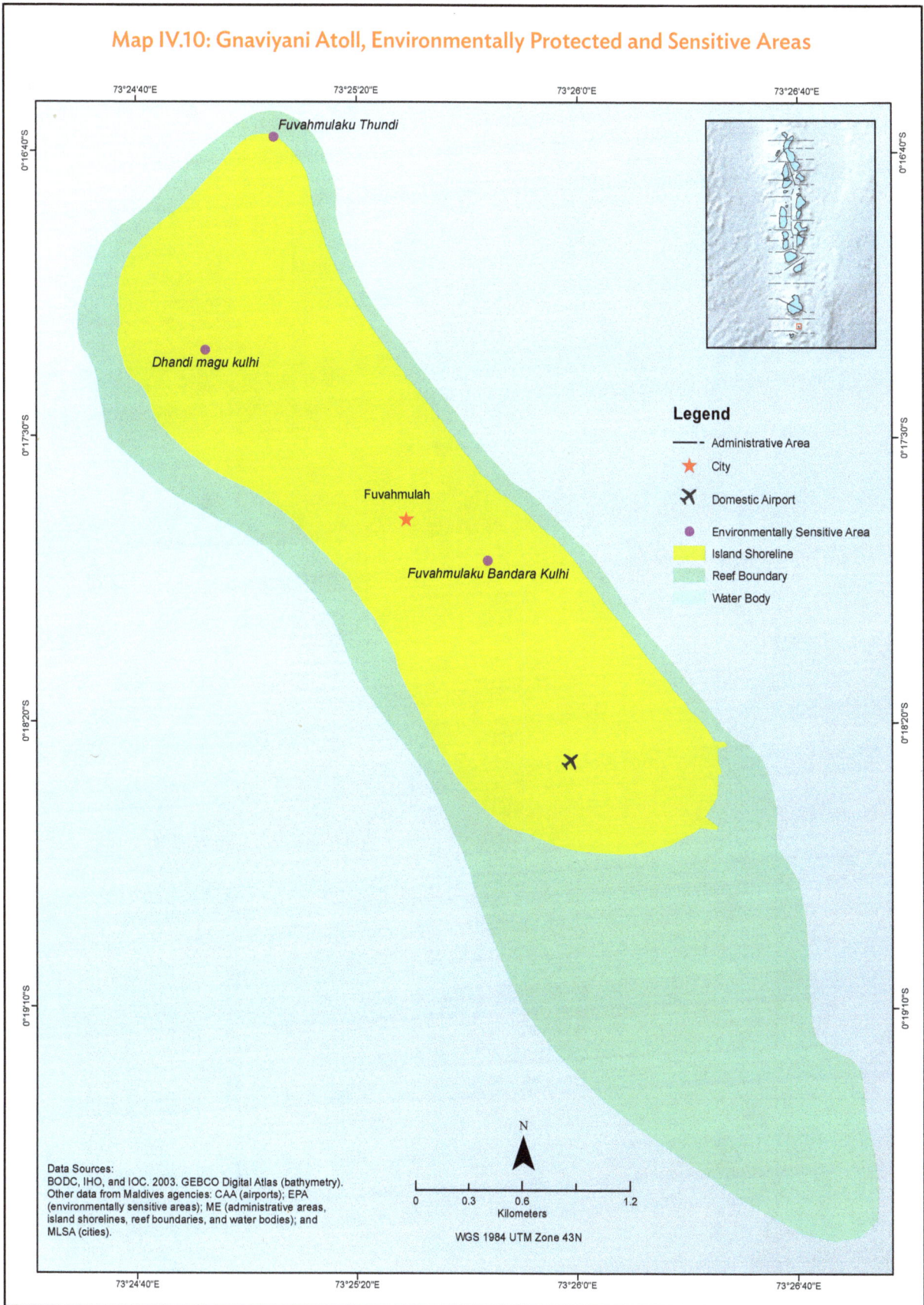

Fuvahmulaku Thundi

Dhandi magu kulhi

Fuvahmulah

Fuvahmulaku Bandara Kulhi

Legend

- - - Administrative Area
- ★ City
- ✈ Domestic Airport
- ● Environmentally Sensitive Area
- Island Shoreline
- Reef Boundary
- Water Body

Data Sources:
BODC, IHO, and IOC. 2003. GEBCO Digital Atlas (bathymetry).
Other data from Maldives agencies: CAA (airports); EPA
(environmentally sensitive areas); ME (administrative areas,
island shorelines, reef boundaries, and water bodies); and
MLSA (cities).

N

| 0 | 0.3 | 0.6 | | 1.2 |

Kilometers

WGS 1984 UTM Zone 43N

Map IV.11: Haa Alifu Atoll, Environmentally Protected and Sensitive Areas

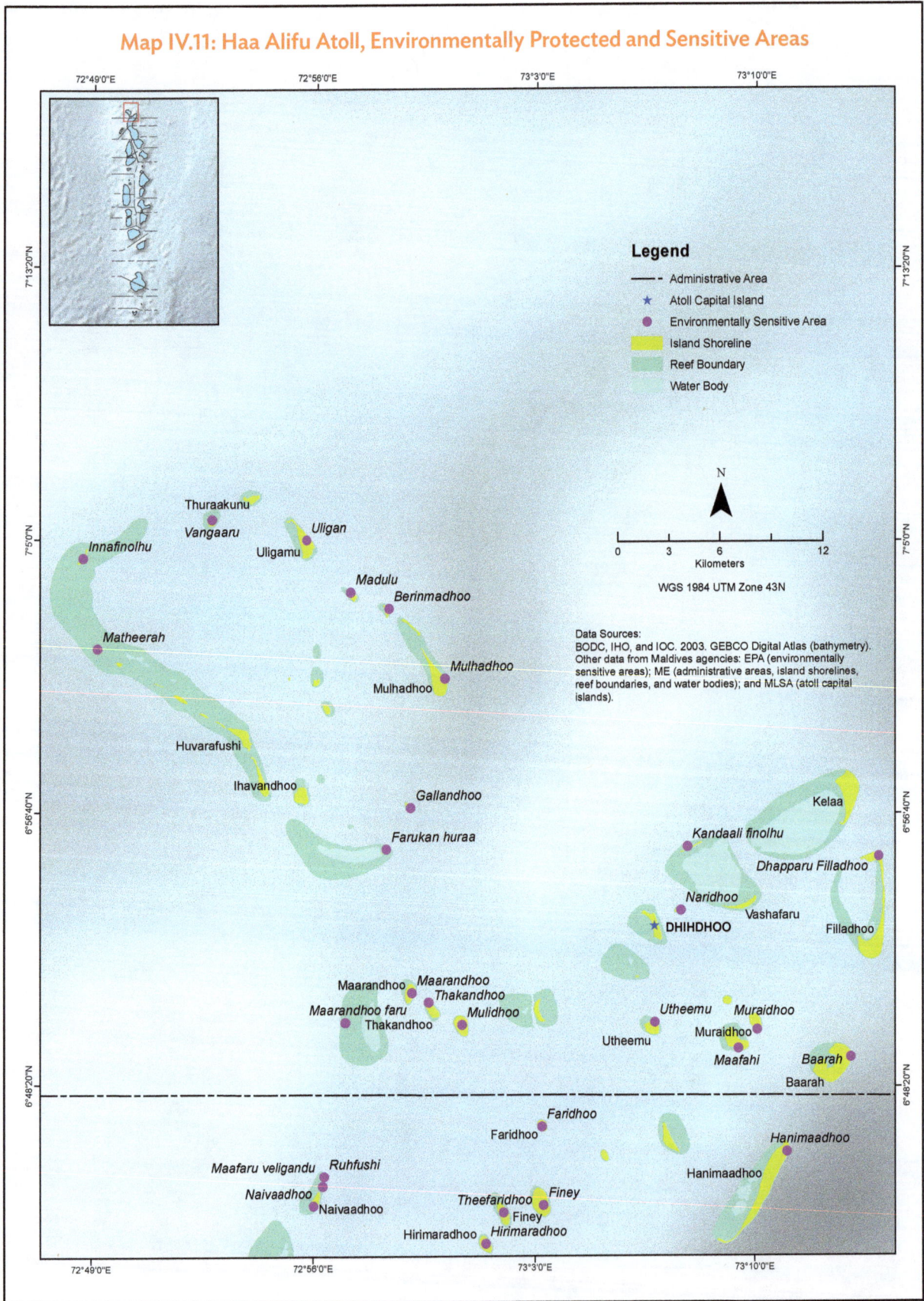

Legend
- – - Administrative Area
- ★ Atoll Capital Island
- ● Environmentally Sensitive Area
- ▮ Island Shoreline
- ▮ Reef Boundary
- ▮ Water Body

N

0   3   6   12
Kilometers

WGS 1984 UTM Zone 43N

Data Sources:
BODC, IHO, and IOC. 2003. GEBCO Digital Atlas (bathymetry). Other data from Maldives agencies: EPA (environmentally sensitive areas); ME (administrative areas, island shorelines, reef boundaries, and water bodies); and MLSA (atoll capital islands).

Thuraakunu
Vangaaru
Uligan
Uligamu
Innafinolhu
Madulu
Berinmadhoo
Matheerah
Mulhadhoo
Mulhadhoo
Huvarafushi
Ihavandhoo
Gallandhoo
Farukan huraa
Kelaa
Kandaali finolhu
Dhapparu Filladhoo
Naridhoo
Vashafaru
Filladhoo
★ DHIHDHOO
Maarandhoo
Maarandhoo
Thakandhoo
Maarandhoo faru
Mulidhoo
Thakandhoo
Utheemu
Muraidhoo
Muraidhoo
Utheemu
Maafahi
Baarah
Baarah
Faridhoo
Faridhoo
Hanimaadhoo
Maafaru veligandu
Ruhfushi
Naivaadhoo
Naivaadhoo
Theefaridhoo
Finey
Finey
Hirimaradhoo
Hirimaradhoo
Hanimaadhoo

Map IV.12: Haa Dhaalu Atoll, Environmentally Protected and Sensitive Areas

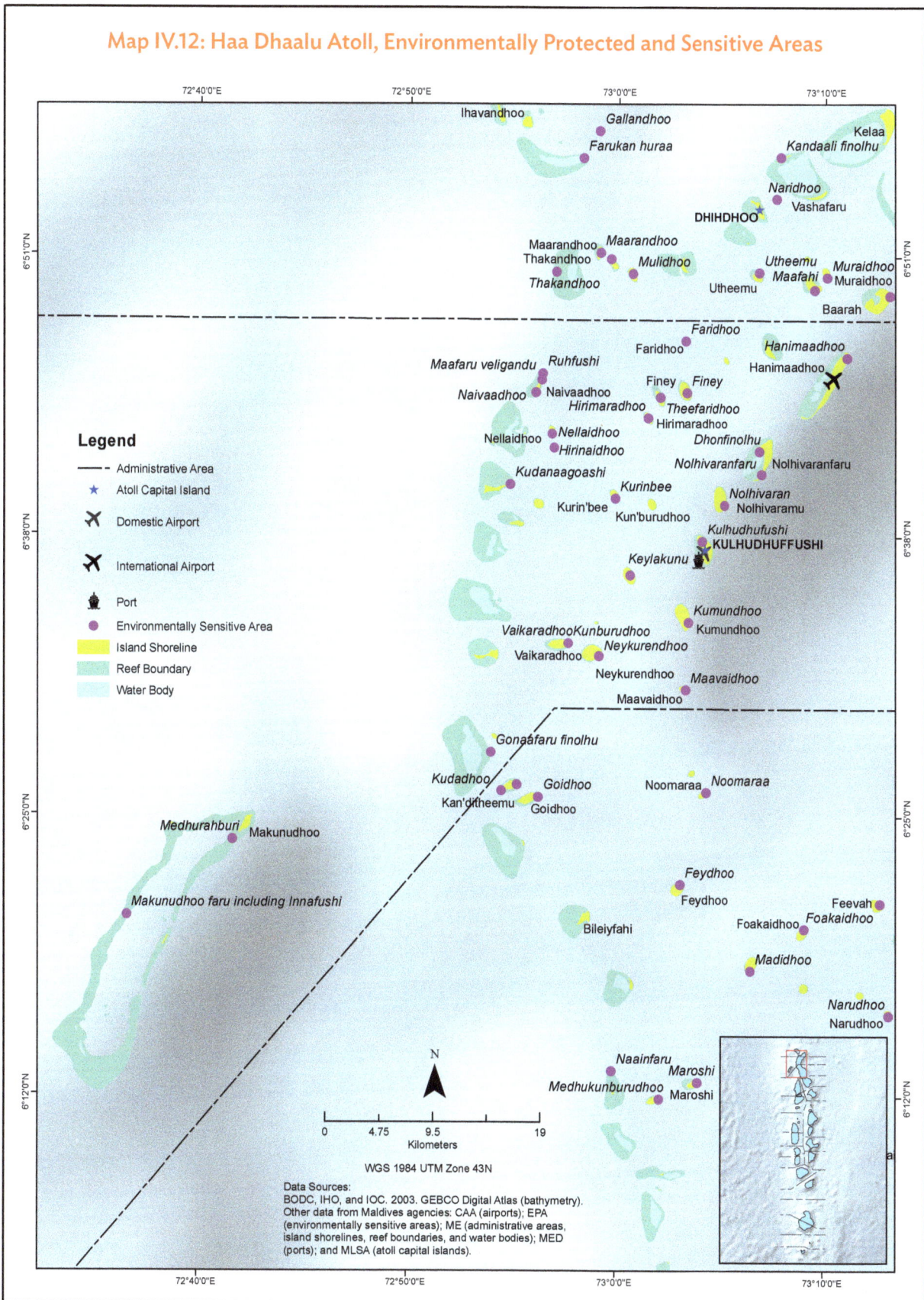

Ihavandhoo
Gallandhoo
Farukan huraa
Kelaa
Kandaali finolhu
Naridhoo
Vashafaru
DHIHDHOO
Maarandhoo
Thakandhoo
Maarandhoo
Mulidhoo
Thakandhoo
Utheemu
Maafahi
Muraidhoo
Muraidhoo
Utheemu
Baarah
Faridhoo
Maafaru veligandu
Ruhfushi
Faridhoo
Hanimaadhoo
Hanimaadhoo
Finey
Finey
Naivaadhoo
Naivaadhoo
Hirimaradhoo
Theefaridhoo
Hirimaradhoo
Nellaidhoo
Nellaidhoo
Dhonfinolhu
Hirinaidhoo
Nolhivaranfaru
Nolhivaranfaru
Kudanaagoashi
Nolhivaran
Kurinbee
Nolhivaramu
Kurin'bee
Kun'burudhoo
Kulhudhufushi
Keylakunu
KULHUDHUFFUSHI
Kumundhoo
Kumundhoo
VaikaradhooKunburudhoo
Neykurendhoo
Vaikaradhoo
Neykurendhoo
Maavaidhoo
Maavaidhoo
Gonaafaru finolhu
Kudadhoo
Goidhoo
Noomaraa
Noomaraa
Kan'ditheemu
Goidhoo
Medhurahburi
Makunudhoo
Feydhoo
Feydhoo
Feevah
Foakaidhoo
Makunudhoo faru including Innafushi
Bileiyfahi
Foakaidhoo
Madidhoo
Narudhoo
Narudhoo
Naainfaru
Maroshi
Medhukunburudhoo
Maroshi

Legend

- Administrative Area
- ★ Atoll Capital Island
- ✈ Domestic Airport
- ✈ International Airport
- ⚓ Port
- ● Environmentally Sensitive Area
- Island Shoreline
- Reef Boundary
- Water Body

72°40'0"E 72°50'0"E 73°0'0"E 73°10'0"E

6°51'0"N
6°38'0"N
6°25'0"N
6°12'0"N

N

0   4.75   9.5   19
Kilometers

WGS 1984 UTM Zone 43N

Data Sources:
BODC, IHO, and IOC. 2003. GEBCO Digital Atlas (bathymetry).
Other data from Maldives agencies: CAA (airports); EPA
(environmentally sensitive areas); ME (administrative areas,
island shorelines, reef boundaries, and water bodies); MED
(ports); and MLSA (atoll capital islands).

Map IV.13: Laamu Atoll, Environmentally Protected and Sensitive Areas

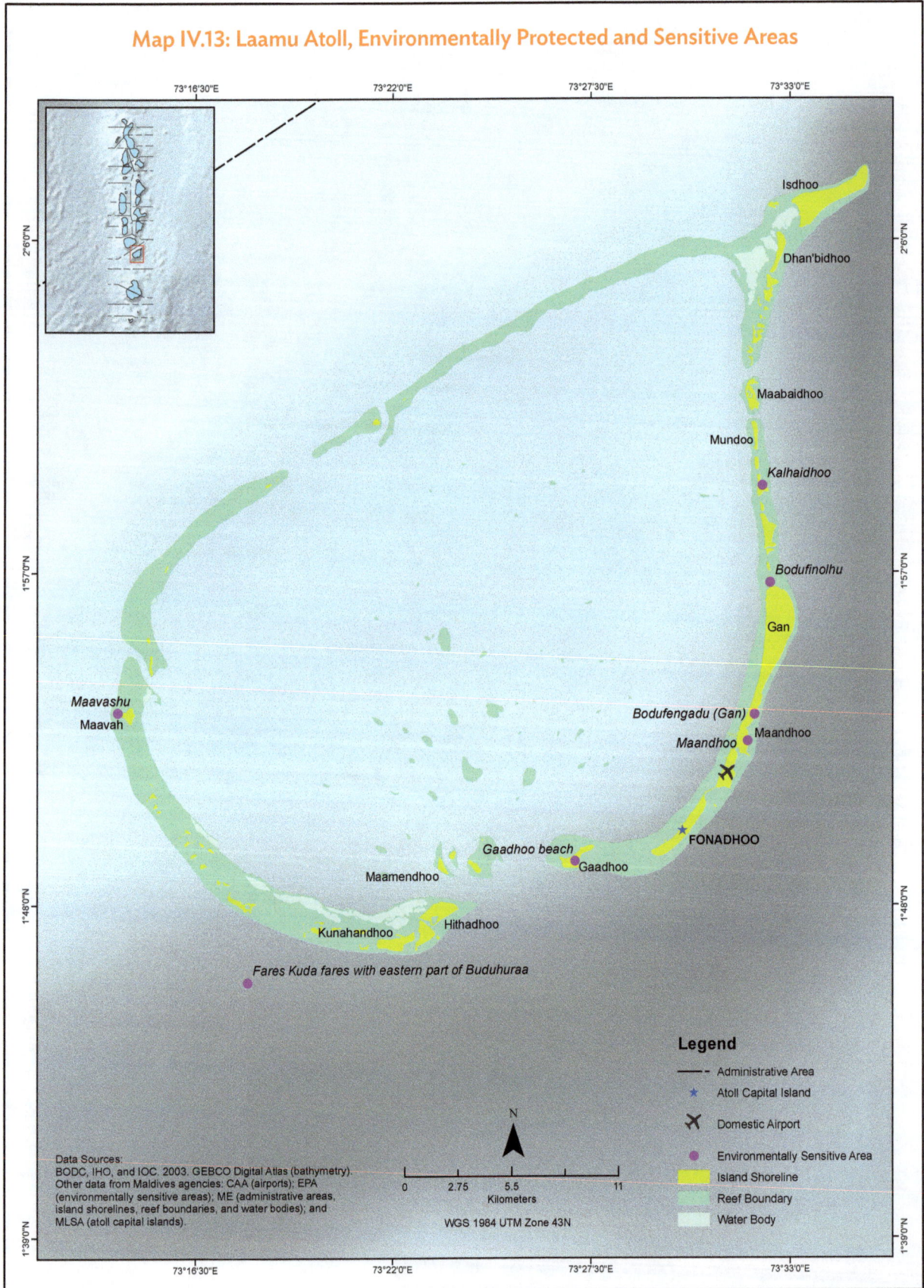

73°16'30"E   73°22'0"E   73°27'30"E   73°33'0"E

2°6'0"N

Isdhoo

Dhan'bidhoo

1°57'0"N

Maabaidhoo

Mundoo

Kalhaidhoo

Bodufinolhu

Gan

Maavashu
Maavah

Bodufengadu (Gan)
Maandhoo
Maandhoo

FONADHOO

Gaadhoo beach
Gaadhoo

Maamendhoo

Kunahandhoo   Hithadhoo

Fares Kuda fares with eastern part of Buduhuraa

1°48'0"N

1°39'0"N

Legend

— — — Administrative Area

★ Atoll Capital Island

✕ Domestic Airport

● Environmentally Sensitive Area

▮ Island Shoreline

▮ Reef Boundary

▮ Water Body

Data Sources:
BODC, IHO, and IOC. 2003. GEBCO Digital Atlas (bathymetry).
Other data from Maldives agencies: CAA (airports); EPA
(environmentally sensitive areas); ME (administrative areas,
island shorelines, reef boundaries, and water bodies); and
MLSA (atoll capital islands).

N

0   2.75   5.5   11
Kilometers

WGS 1984 UTM Zone 43N

Map IV.14: Lhaviyani Atoll, Environmentally Protected and Sensitive Areas

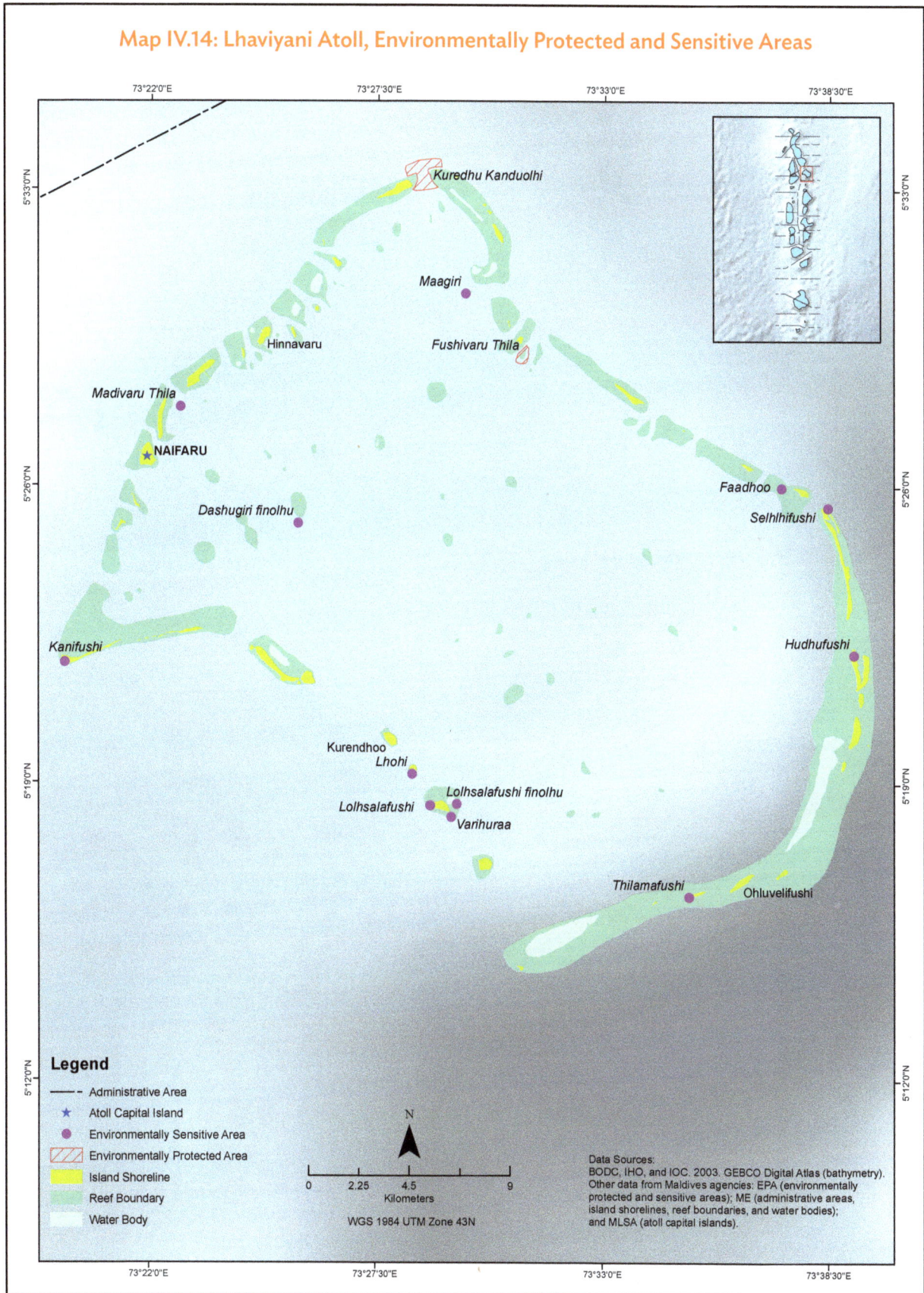

73°22'0"E   73°27'30"E   73°33'0"E   73°38'30"E

5°33'0"N

Kuredhu Kanduolhi

Maagiri

Hinnavaru

Fushivaru Thila

Madivaru Thila

★ NAIFARU

5°26'0"N

Faadhoo

Selhlhifushi

Dashugiri finolhu

Hudhufushi

Kanifushi

Kurendhoo

Lhohi

Lolhsalafushi finolhu

Lolhsalafushi

Varihuraa

Thilamafushi

Ohluvelifushi

5°19'0"N

5°12'0"N

Legend

— Administrative Area
★ Atoll Capital Island
● Environmentally Sensitive Area
▨ Environmentally Protected Area
▮ Island Shoreline
▮ Reef Boundary
▮ Water Body

N

0   2.25   4.5   9
Kilometers

WGS 1984 UTM Zone 43N

Data Sources:
BODC, IHO, and IOC. 2003. GEBCO Digital Atlas (bathymetry).
Other data from Maldives agencies: EPA (environmentally
protected and sensitive areas); ME (administrative areas,
island shorelines, reef boundaries, and water bodies);
and MLSA (atoll capital islands).

73°22'0"E   73°27'30"E   73°33'0"E   73°38'30"E

Map IV.15: Meemu Atoll, Environmentally Protected and Sensitive Areas

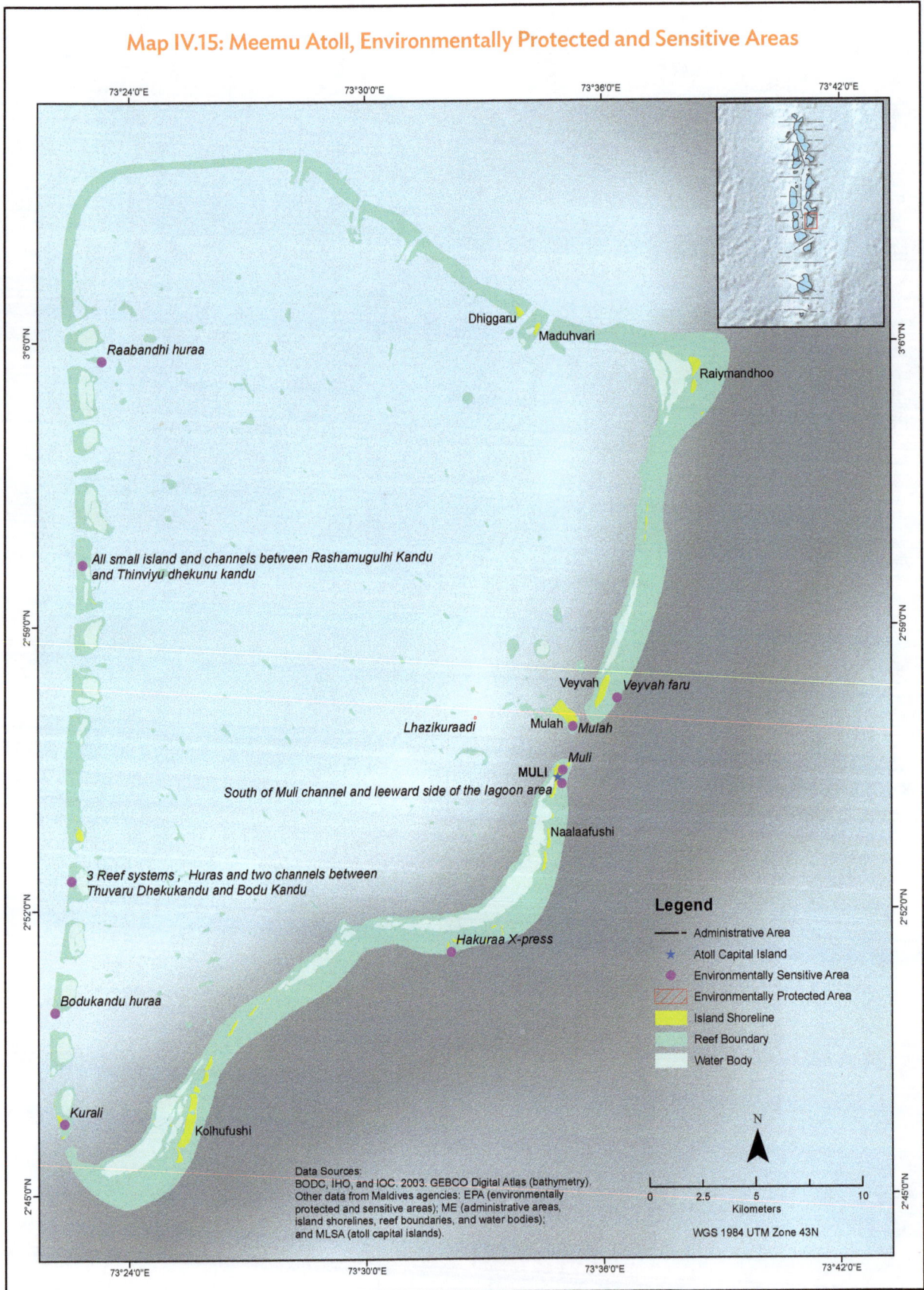

Dhiggaru

Maduhvari

Raiymandhoo

Raabandhi huraa

All small island and channels between Rashamugulhi Kandu and Thinviyu dhekunu kandu

Veyvah   *Veyvah faru*

Lhazikuraadi   Mulah   *Mulah*

Muli

MULI

South of Muli channel and leeward side of the lagoon area

Naalaafushi

3 Reef systems , Huras and two channels between Thuvaru Dhekukandu and Bodu Kandu

Hakuraa X-press

Bodukandu huraa

Kurali   Kolhufushi

Legend

— – — Administrative Area

★ Atoll Capital Island

● Environmentally Sensitive Area

▨ Environmentally Protected Area

▉ Island Shoreline

▉ Reef Boundary

▉ Water Body

N

Data Sources:
BODC, IHO, and IOC. 2003. GEBCO Digital Atlas (bathymetry).
Other data from Maldives agencies: EPA (environmentally
protected and sensitive areas); ME (administrative areas,
island shorelines, reef boundaries, and water bodies);
and MLSA (atoll capital islands).

0   2.5   5   10
Kilometers

WGS 1984 UTM Zone 43N

Map IV.16: Noonu Atoll, Environmentally Protected and Sensitive Areas

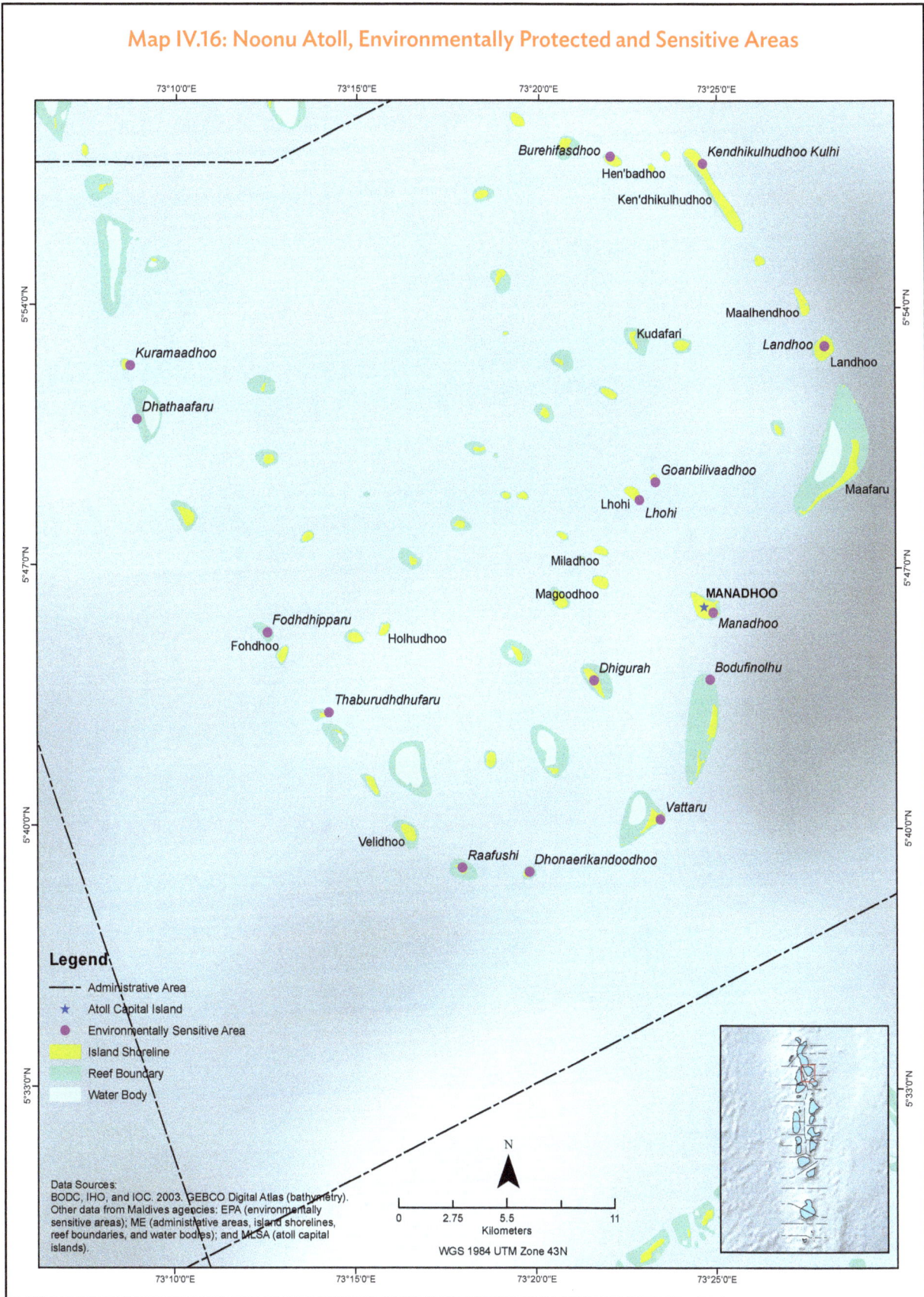

Legend

- Administrative Area
- ★ Atoll Capital Island
- ● Environmentally Sensitive Area
- Island Shoreline
- Reef Boundary
- Water Body

Data Sources:
BODC, IHO, and IOC. 2003. GEBCO Digital Atlas (bathymetry).
Other data from Maldives agencies: EPA (environmentally sensitive areas); ME (administrative areas, island shorelines, reef boundaries, and water bodies); and MLSA (atoll capital islands).

N

0   2.75   5.5   11
Kilometers

WGS 1984 UTM Zone 43N

Map IV.17: North Malé Atoll, Environmentally Protected and Sensitive Areas

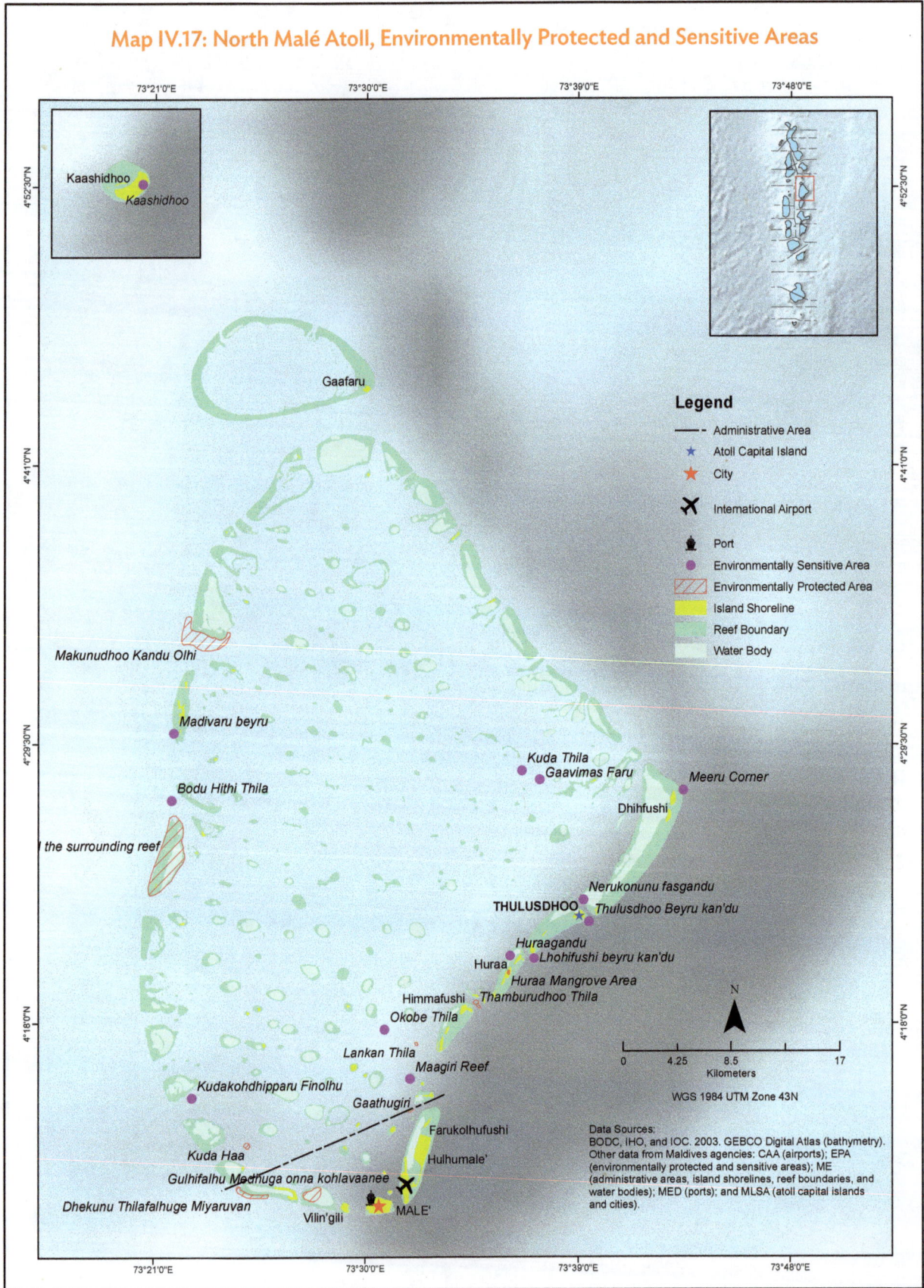

Kaashidhoo
Kaashidhoo

Gaafaru

Legend

- ----- Administrative Area
- ⭑ Atoll Capital Island
- ★ City
- ✈ International Airport
- ⚓ Port
- ● Environmentally Sensitive Area
- ▨ Environmentally Protected Area
- ▨ Island Shoreline
- Reef Boundary
- Water Body

Makunudhoo Kandu Olhi

● *Madivaru beyru*

Kuda Thila
Gaavimas Faru
● *Meeru Corner*

● *Bodu Hithi Thila*

Dhihfushi

I the surrounding reef

Nerukonunu fasgandu
THULUSDHOO ⭑ *Thulusdhoo Beyru kan'du*

Huraagandu
● *Lhohifushi beyru kan'du*
Huraa ●
Huraa Mangrove Area
Himmafushi   *Thamburudhoo Thila*
● *Okobe Thila*

Lankan Thila
● *Maagiri Reef*

● *Kudakohdhipparu Finolhu*
Gaathugiri

Farukolhufushi
Hulhumale'

N
▲

| 0 | 4.25 | 8.5 | | 17 |
Kilometers

WGS 1984 UTM Zone 43N

Kuda Haa
Gulhifalhu Medhuga onna kohlavaanee   ✈
Dhekunu Thilafalhuge Miyaruvan
Vilin'gili   ★ MALE'

Data Sources:
BODC, IHO, and IOC. 2003. GEBCO Digital Atlas (bathymetry).
Other data from Maldives agencies: CAA (airports); EPA
(environmentally protected and sensitive areas); ME
(administrative areas, island shorelines, reef boundaries, and
water bodies); MED (ports); and MLSA (atoll capital islands
and cities).

Map IV.18: Raa Atoll, Environmentally Protected and Sensitive Areas

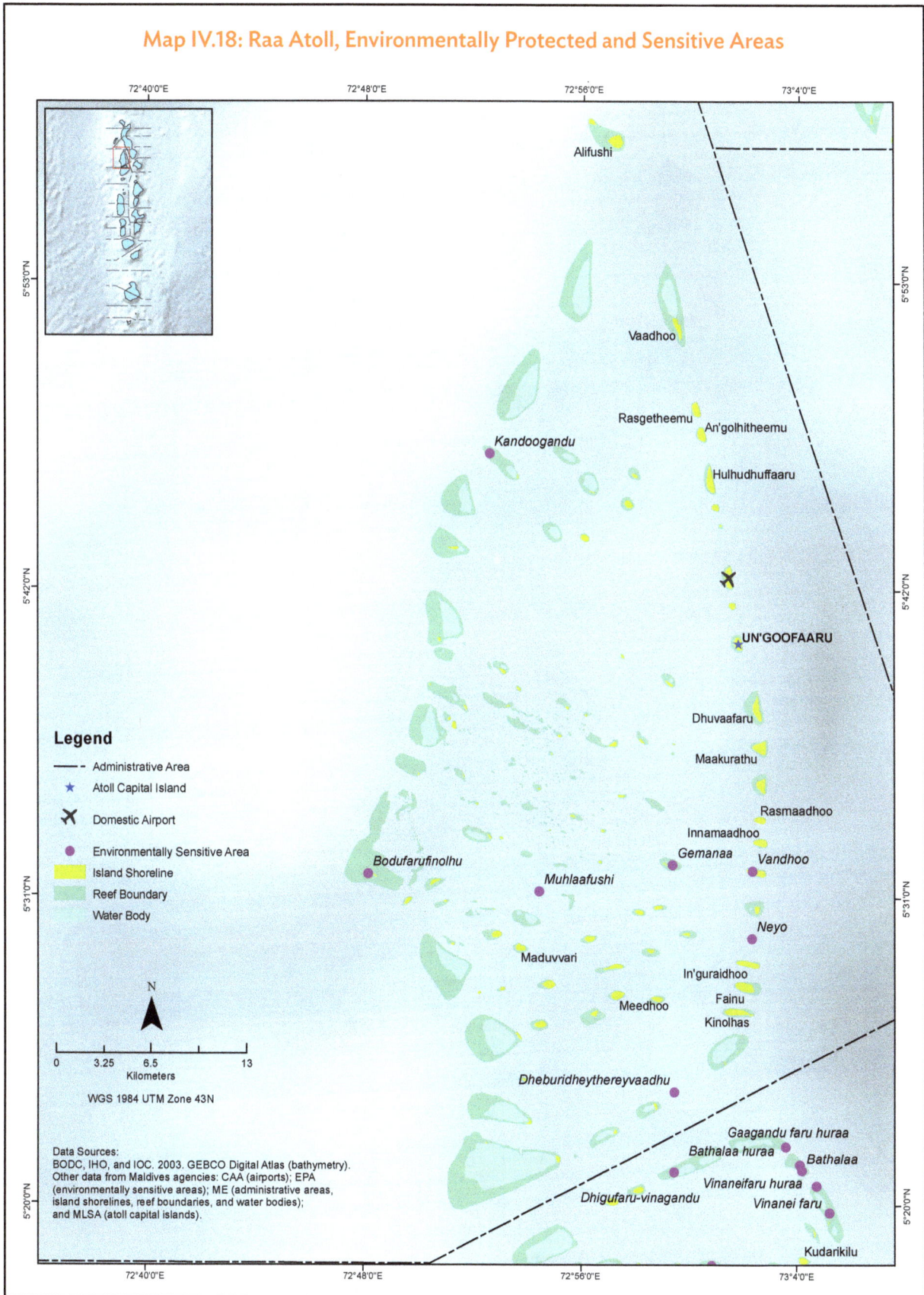

72°40'0"E   72°48'0"E   72°56'0"E   73°4'0"E

Alifushi

Vaadhoo

Rasgetheemu   An'golhitheemu

Hulhudhuffaaru

Kandoogandu

5°42'0"N

UN'GOOFAARU

Dhuvaafaru

Maakurathu

Rasmaadhoo

Innamaadhoo

Gemanaa   *Vandhoo*

Bodufarufinolhu

Muhlaafushi

Neyo

In'guraidhoo

Fainu

Maduvvari

Meedhoo   Kinolhas

Dheburidheythereyvaadhu

Gaagandu faru huraa

Bathalaa huraa   *Bathalaa*

Vinaneifaru huraa

Dhigufaru-vinagandu   *Vinanei faru*

Kudarikilu

Legend

- — — Administrative Area
- ★ Atoll Capital Island
- ✈ Domestic Airport
- ● Environmentally Sensitive Area
- ▮ Island Shoreline
- ▮ Reef Boundary
- Water Body

N

0   3.25   6.5   13
Kilometers

WGS 1984 UTM Zone 43N

Data Sources:
BODC, IHO, and IOC. 2003. GEBCO Digital Atlas (bathymetry).
Other data from Maldives agencies: CAA (airports); EPA
(environmentally sensitive areas); ME (administrative areas,
island shorelines, reef boundaries, and water bodies);
and MLSA (atoll capital islands).

Map IV.19: Shaviyani Atoll, Environmentally Protected and Sensitive Areas

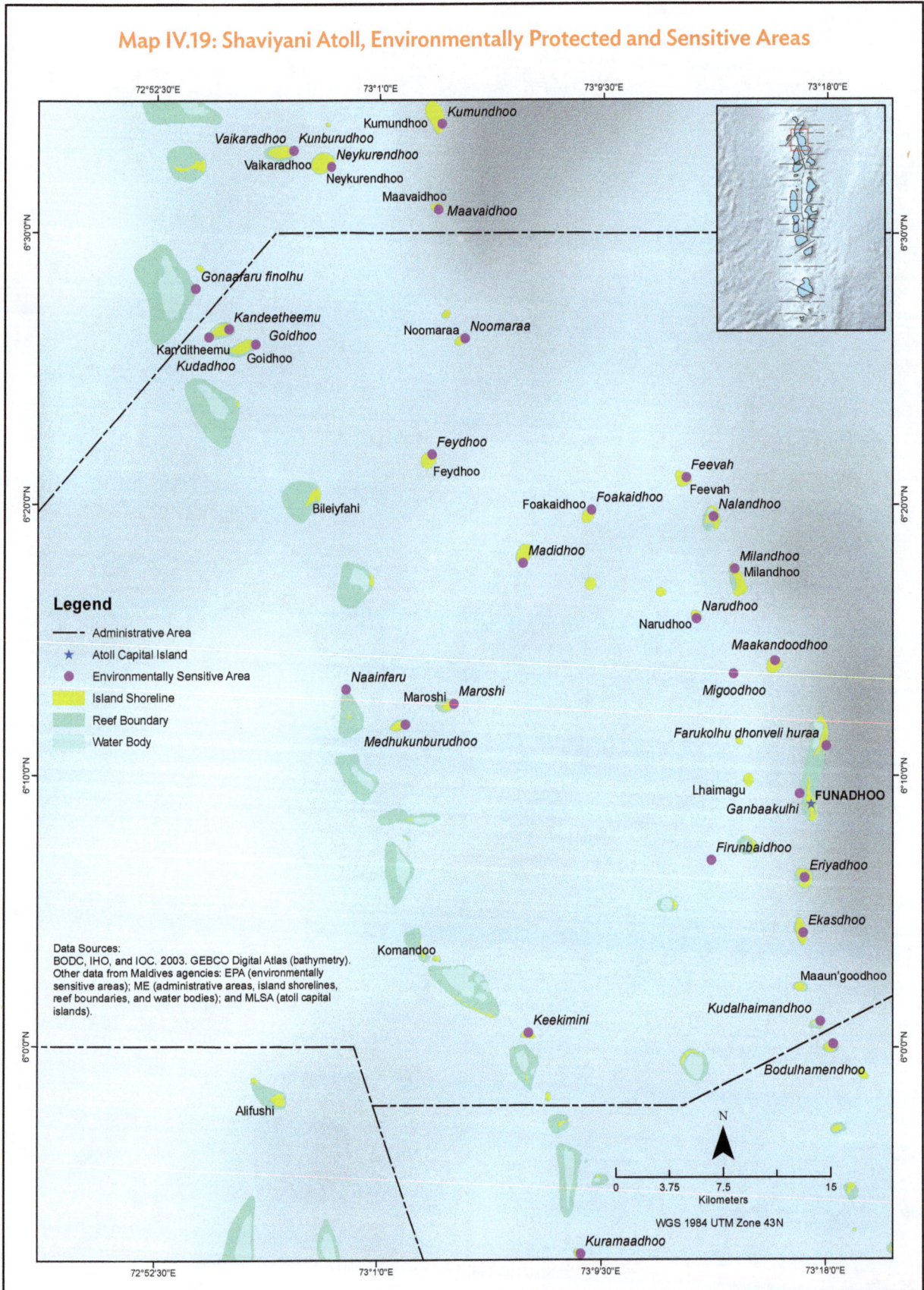

Kumundhoo
Kumundhoo

Vaikaradhoo   Kunburudhoo
Vaikaradhoo   Neykurendhoo
Neykurendhoo

Maavaidhoo
Maavaidhoo

Gonaafaru finolhu

Kandeetheemu
Goidhoo
Kanditheemu
Kudadhoo   Goidhoo

Noomaraa   Noomaraa

Feydhoo
Feydhoo

Feevah
Feevah

Foakaidhoo   Foakaidhoo
Nalandhoo

Bileiyfahi

Madidhoo

Milandhoo
Milandhoo

Narudhoo
Narudhoo

Maakandoodhoo

Legend

- - - Administrative Area
★   Atoll Capital Island
●   Environmentally Sensitive Area
▮   Island Shoreline
▮   Reef Boundary
▮   Water Body

Naainfaru
Maroshi   Maroshi

Migoodhoo

Medhukunburudhoo

Farukolhu dhonveli huraa

Lhaimagu   FUNADHOO
Ganbaakulhi

Firunbaidhoo

Eriyadhoo

Data Sources:
BODC, IHO, and IOC. 2003. GEBCO Digital Atlas (bathymetry).
Other data from Maldives agencies: EPA (environmentally
sensitive areas); ME (administrative areas, island shorelines,
reef boundaries, and water bodies); and MLSA (atoll capital
islands).

Ekasdhoo

Komandoo

Maaun'goodhoo

Kudalhaimandhoo

Keekimini

Bodulhamendhoo

Alifushi

N

0   3.75   7.5   15
Kilometers

WGS 1984 UTM Zone 43N

Kuramaadhoo

Map IV.20: South Malé Atoll, Environmentally Protected and Sensitive Areas

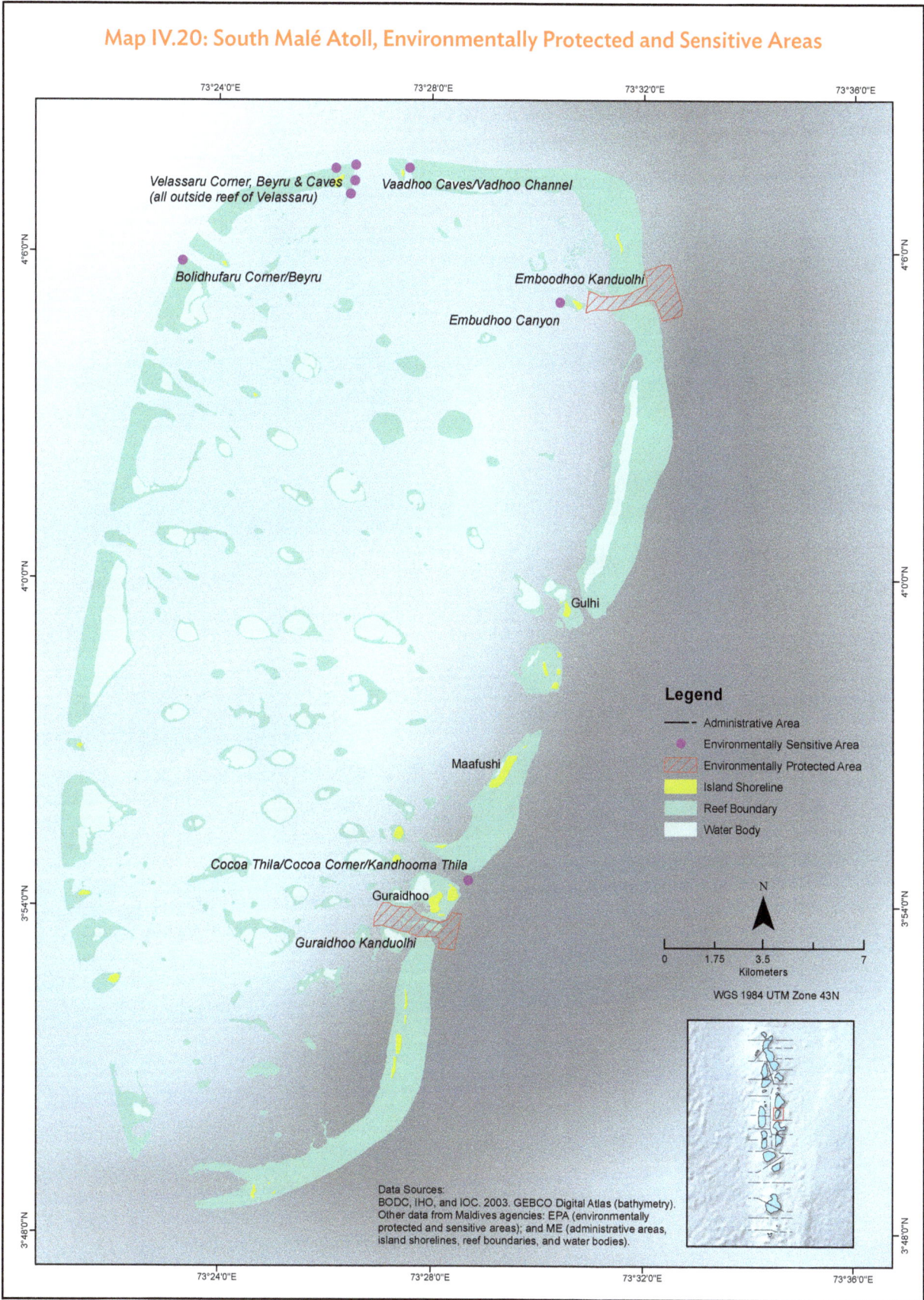

73°24'0"E   73°28'0"E   73°32'0"E   73°36'0"E

4°6'0"N

Velassaru Corner, Beyru & Caves
(all outside reef of Velassaru)

Vaadhoo Caves/Vadhoo Channel

Bolidhufaru Corner/Beyru

Emboodhoo Kanduolhi

Embudhoo Canyon

4°0'0"N

Gulhi

Legend

— - Administrative Area
• Environmentally Sensitive Area
▨ Environmentally Protected Area
▮ Island Shoreline
▮ Reef Boundary
▯ Water Body

Maafushi

3°54'0"N

Cocoa Thila/Cocoa Corner/Kandhooma Thila

Guraidhoo

Guraidhoo Kanduolhi

N

0   1.75   3.5   7
Kilometers

WGS 1984 UTM Zone 43N

3°48'0"N

Data Sources:
BODC, IHO, and IOC. 2003. GEBCO Digital Atlas (bathymetry).
Other data from Maldives agencies: EPA (environmentally
protected and sensitive areas); and ME (administrative areas,
island shorelines, reef boundaries, and water bodies).

73°24'0"E   73°28'0"E   73°32'0"E   73°36'0"E

Map IV.21: Thaa Atoll, Environmentally Protected and Sensitive Areas

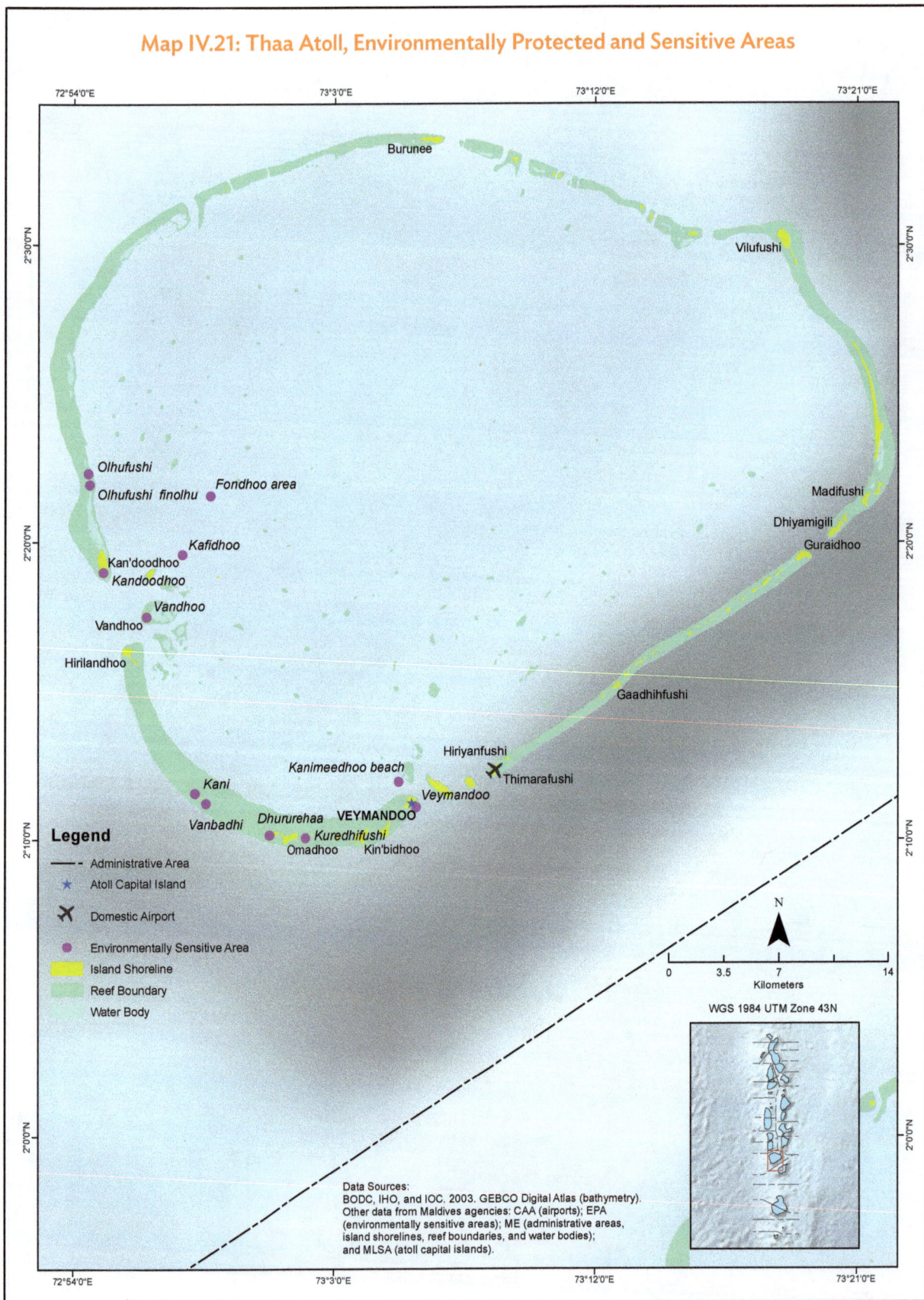

Burunee

Vilufushi

Olhufushi
Olhufushi finolhu   Fondhoo area

Madifushi
Dhiyamigili
Kafidhoo   Guraidhoo
Kan'doodhoo
Kandoodhoo
Vandhoo
Vandhoo

Hirilandhoo

Gaadhihfushi

Hiriyanfushi
Kanimeedhoo beach   Thimarafushi
Kani   Veymandoo
Vanbadhi   Dhururehaa   VEYMANDOO
Omadhoo   Kuredhifushi
Kin'bidhoo

Legend

— Administrative Area
★ Atoll Capital Island
✕ Domestic Airport
● Environmentally Sensitive Area
▇ Island Shoreline
▇ Reef Boundary
▇ Water Body

N

0   3.5   7   14
Kilometers

WGS 1984 UTM Zone 43N

Data Sources:
BODC, IHO, and IOC. 2003. GEBCO Digital Atlas (bathymetry).
Other data from Maldives agencies: CAA (airports); EPA
(environmentally sensitive areas); ME (administrative areas,
island shorelines, reef boundaries, and water bodies);
and MLSA (atoll capital islands).

Map IV.22: Vaavu Atoll, Environmentally Protected and Sensitive Areas

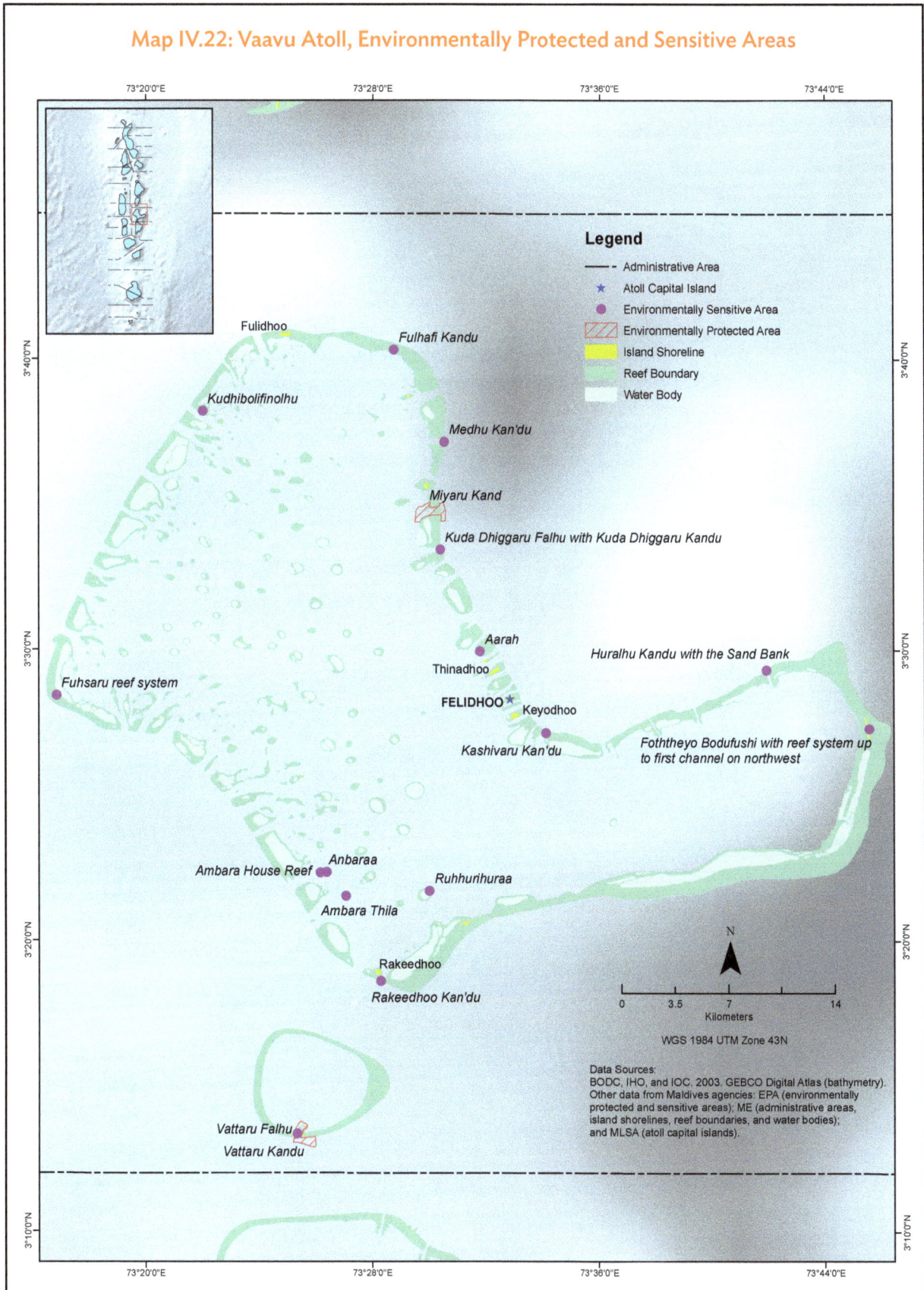

73°20'0"E   73°28'0"E   73°36'0"E   73°44'0"E

Legend

- — — Administrative Area
- ★ Atoll Capital Island
- ● Environmentally Sensitive Area
- ▨ Environmentally Protected Area
- ▉ Island Shoreline
- Reef Boundary
- Water Body

Fulidhoo

Fulhafi Kandu

Kudhibolifinolhu

Medhu Kan'du

Miyaru Kand

Kuda Dhiggaru Falhu with Kuda Dhiggaru Kandu

Aarah

Thinadhoo

Huralhu Kandu with the Sand Bank

Fuhsaru reef system

FELIDHOO ★

Keyodhoo

Foththeyo Bodufushi with reef system up to first channel on northwest

Kashivaru Kan'du

Anbaraa

Ambara House Reef

Ruhhurihuraa

Ambara Thila

Rakeedhoo

N

Rakeedhoo Kan'du

0   3.5   7   14
Kilometers

WGS 1984 UTM Zone 43N

Data Sources:
BODC, IHO, and IOC. 2003. GEBCO Digital Atlas (bathymetry).
Other data from Maldives agencies: EPA (environmentally
protected and sensitive areas); ME (administrative areas,
island shorelines, reef boundaries, and water bodies);
and MLSA (atoll capital islands).

Vattaru Falhu
Vattaru Kandu

3°40'0"N   3°30'0"N   3°20'0"N   3°10'0"N

73°20'0"E   73°28'0"E   73°36'0"E   73°44'0"E

Coastal Protection

Ocean waves, the changing mean sea level, beach erosion, and anthropogenic activities such as coral and sand mining threaten the coastal ecosystems. To protect their precious resources, the people of Maldives have identified sites to be safeguarded against coastal erosion. The coasts of 51 inhabited islands have coastal protection sites. Coastal protection is an important task for Maldivians as the coastal ecosystems—specifically the mangroves and coral reefs—also serve as a buffer from storm surges, inundation, and tsunamis. Having a well-protected coast also preserves the country's rich aquatic resources.

Table IV.1: Maldives, Islands with Coastal Protection

Atoll	Number	Island	Atoll	Number	Island
Addu City	1	Feydhoo	Laamu	1	Gaadhoo
Alifu Alifu	3	Bodufolhudhoo	Lhaviyani	2	Hinnavaru
		Rasdhoo[a]			Kurehdhoo
		Ukulhas	Malé City	5	Hulhule
Alifu Dhaalu	2	Kun'burudhoo			HulhuMalé
		Maamigili			Malé[b]
Baa	3	Eydhafushi[a]			Thilafushi
		Fares			Vilin'gili
		Thulhaadhoo	Meemu	3	Dhiggaru
Dhaalu	4	Kudahuvadhoo[a]			Mulia
		Maaen'boodhoo			Naalaafushi
		Meedhoo	Noonu	2	Holhudhoo
		Rin'budhoo			Maafaru
Faafu	1	Nilandhoo[a]	Raa	3	Dhuvaafaru
Gaafu Alifu	3	Vilin'gili[a]			Fainu
		Faresmaathodaa			Maduvvari
		Gahdhoo	Shaviyani	2	Bileiyfahi
Haa Alifu	3	Dhidhdhoo[a]			Komandoo
		Ihavandhoo	Thaa	6	Dhiyamigili
		Kelaa			Guraidhoo

continued on next page

Table IV.1 *continued*

Atoll	Number	Island	Atoll	Number	Island
Haa Dhaalu	3	Finey			Kan'doodhoo
		Kulhudhuffushi[a]			Madifushi
		Neykurendhoo			Thimarafushi
Kaafu	4	Guraidhoo			Vilufushi
		Himmafushi			
		Maafushi			
		Thulusdhoo[a]			

Notes:
[a] Atoll capital.
[b] City.

Source: Ministry of National Planning and Infrastructure, 2017.

Map IV.23: Maldives, Coastal Protection

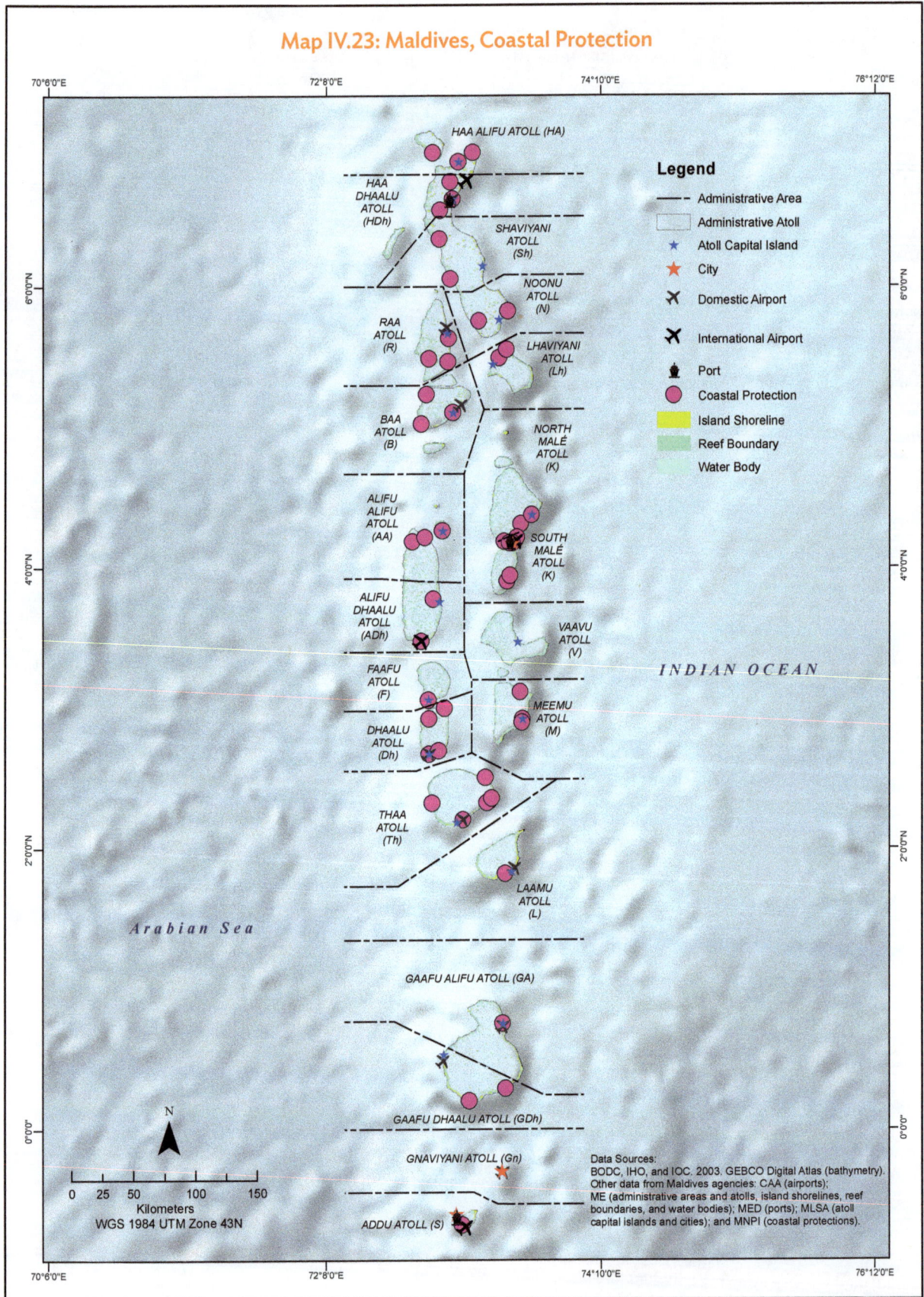

Legend

- — — Administrative Area
- ☐ Administrative Atoll
- ★ Atoll Capital Island
- ★ City
- ✕ Domestic Airport
- ✈ International Airport
- ⚓ Port
- ● Coastal Protection
- ▮ Island Shoreline
- ▮ Reef Boundary
- ▮ Water Body

HAA ALIFU ATOLL (HA)

HAA DHAALU ATOLL (HDh)

SHAVIYANI ATOLL (Sh)

NOONU ATOLL (N)

RAA ATOLL (R)

LHAVIYANI ATOLL (Lh)

BAA ATOLL (B)

NORTH MALÉ ATOLL (K)

ALIFU ALIFU ATOLL (AA)

SOUTH MALÉ ATOLL (K)

ALIFU DHAALU ATOLL (ADh)

VAAVU ATOLL (V)

INDIAN OCEAN

FAAFU ATOLL (F)

MEEMU ATOLL (M)

DHAALU ATOLL (Dh)

THAA ATOLL (Th)

LAAMU ATOLL (L)

Arabian Sea

GAAFU ALIFU ATOLL (GA)

GAAFU DHAALU ATOLL (GDh)

GNAVIYANI ATOLL (Gn)

ADDU ATOLL (S)

N

0   25   50   100   150
Kilometers
WGS 1984 UTM Zone 43N

Data Sources:
BODC, IHO, and IOC. 2003. GEBCO Digital Atlas (bathymetry).
Other data from Maldives agencies: CAA (airports);
ME (administrative areas and atolls, island shorelines, reef
boundaries, and water bodies); MED (ports); MLSA (atoll
capital islands and cities); and MNPI (coastal protections).

70°6'0"E   72°8'0"E   74°10'0"E   76°12'0"E

6°0'0"N

4°0'0"N

2°0'0"N

0°0'0"

Map IV.24: Addu City, Coastal Protection

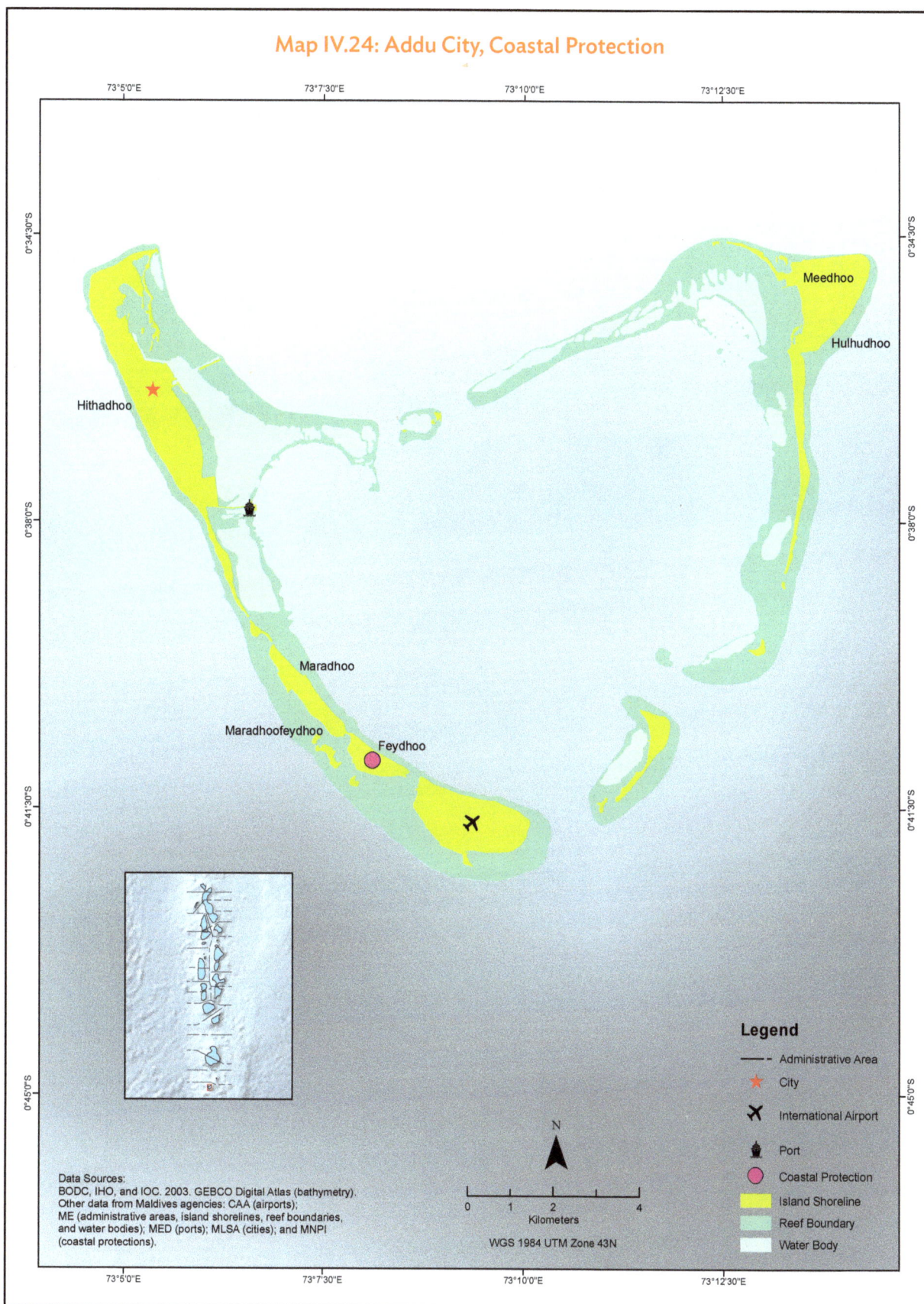

73°5'0"E   73°7'30"E   73°10'0"E   73°12'30"E

0°34'30"S   0°34'30"S
0°38'0"S   0°38'0"S
0°41'30"S   0°41'30"S
0°45'0"S   0°45'0"S

Meedhoo
Hulhudhoo
Hithadhoo
Maradhoo
Maradhoofeydhoo
Feydhoo

Legend
- - - Administrative Area
★ City
✈ International Airport
⚓ Port
● Coastal Protection
Island Shoreline
Reef Boundary
Water Body

N

0 1 2 4
Kilometers
WGS 1984 UTM Zone 43N

Data Sources:
BODC, IHO, and IOC. 2003. GEBCO Digital Atlas (bathymetry).
Other data from Maldives agencies: CAA (airports);
ME (administrative areas, island shorelines, reef boundaries,
and water bodies); MED (ports); MLSA (cities); and MNPI
(coastal protections).

Map IV.25: Alifu Alifu Atoll, Coastal Protection

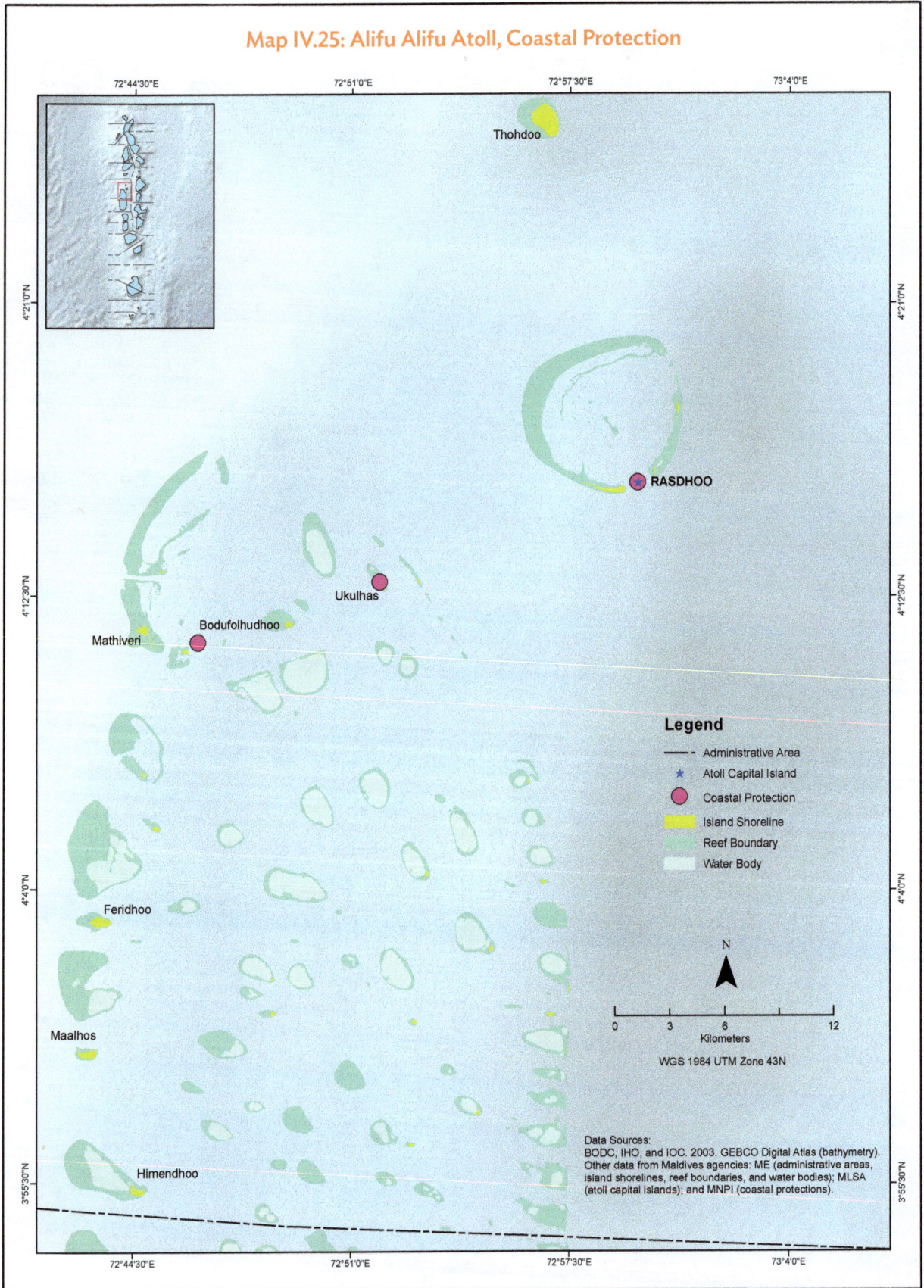

Thohdoo

RASDHOO

Ukulhas

Bodufolhudhoo

Mathiveri

Feridhoo

Maalhos

Himendhoo

Legend

- – – Administrative Area
- ★ Atoll Capital Island
- ⬤ Coastal Protection
- Island Shoreline
- Reef Boundary
- Water Body

N

0   3   6   12
Kilometers

WGS 1984 UTM Zone 43N

Data Sources:
BODC, IHO, and IOC. 2003. GEBCO Digital Atlas (bathymetry).
Other data from Maldives agencies: ME (administrative areas,
island shorelines, reef boundaries, and water bodies); MLSA
(atoll capital islands); and MNPI (coastal protections).

Map IV.26: Alifu Dhaalu Atoll, Coastal Protection

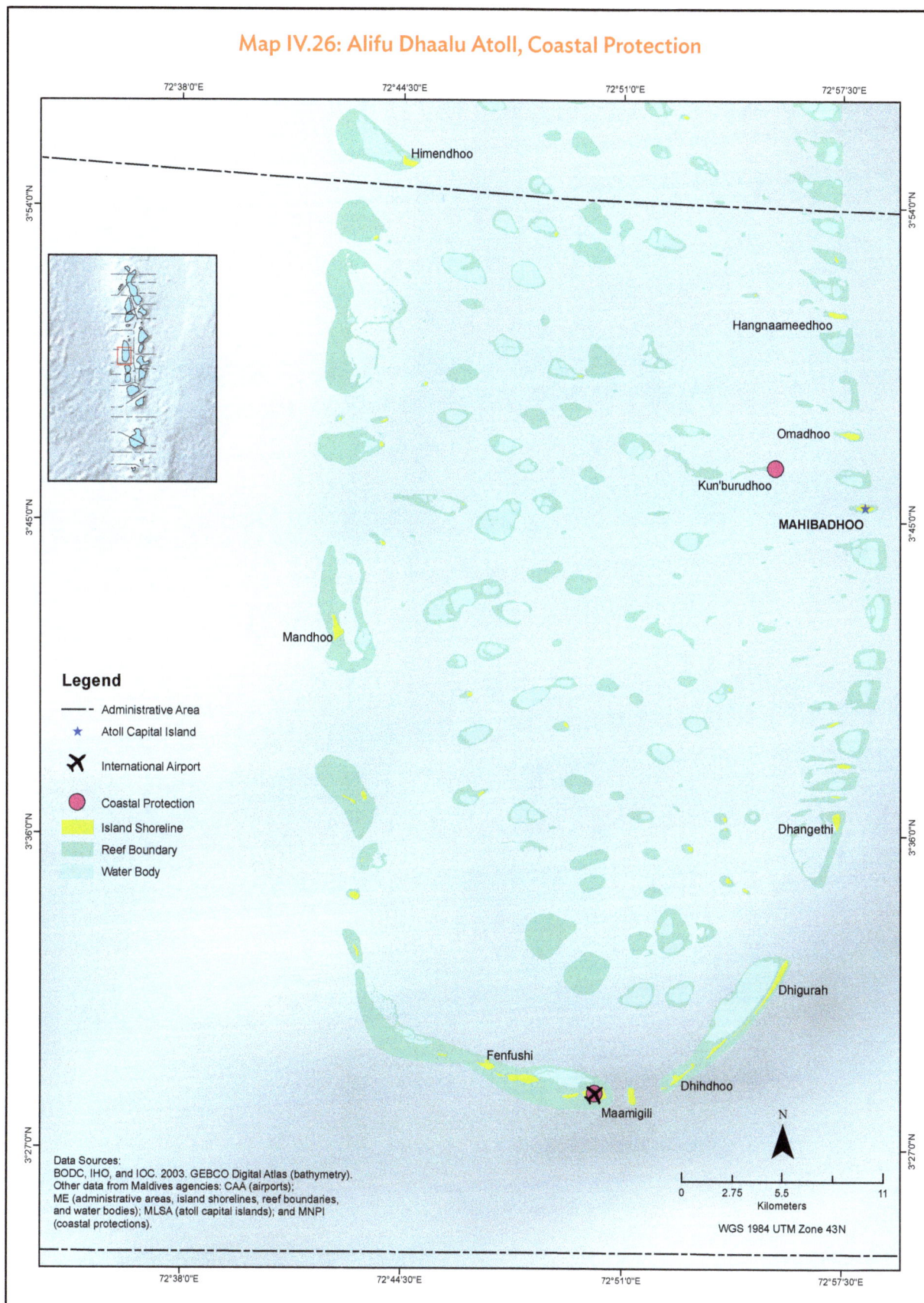

Himendhoo

Hangnaameedhoo

Omadhoo

Kun'burudhoo

MAHIBADHOO

Mandhoo

Dhangethi

Dhigurah

Fenfushi

Dhihdhoo

Maamigili

Legend

— — — Administrative Area

★ Atoll Capital Island

✈ International Airport

● Coastal Protection

  Island Shoreline

  Reef Boundary

  Water Body

N

0   2.75   5.5   11
Kilometers

WGS 1984 UTM Zone 43N

Data Sources:
BODC, IHO, and IOC. 2003. GEBCO Digital Atlas (bathymetry).
Other data from Maldives agencies: CAA (airports);
ME (administrative areas, island shorelines, reef boundaries,
and water bodies); MLSA (atoll capital islands); and MNPI
(coastal protections).

72°38'0"E   72°44'30"E   72°51'0"E   72°57'30"E

3°54'0"N
3°45'0"N
3°36'0"N
3°27'0"N

Map IV.27: Baa Atoll, Coastal Protection

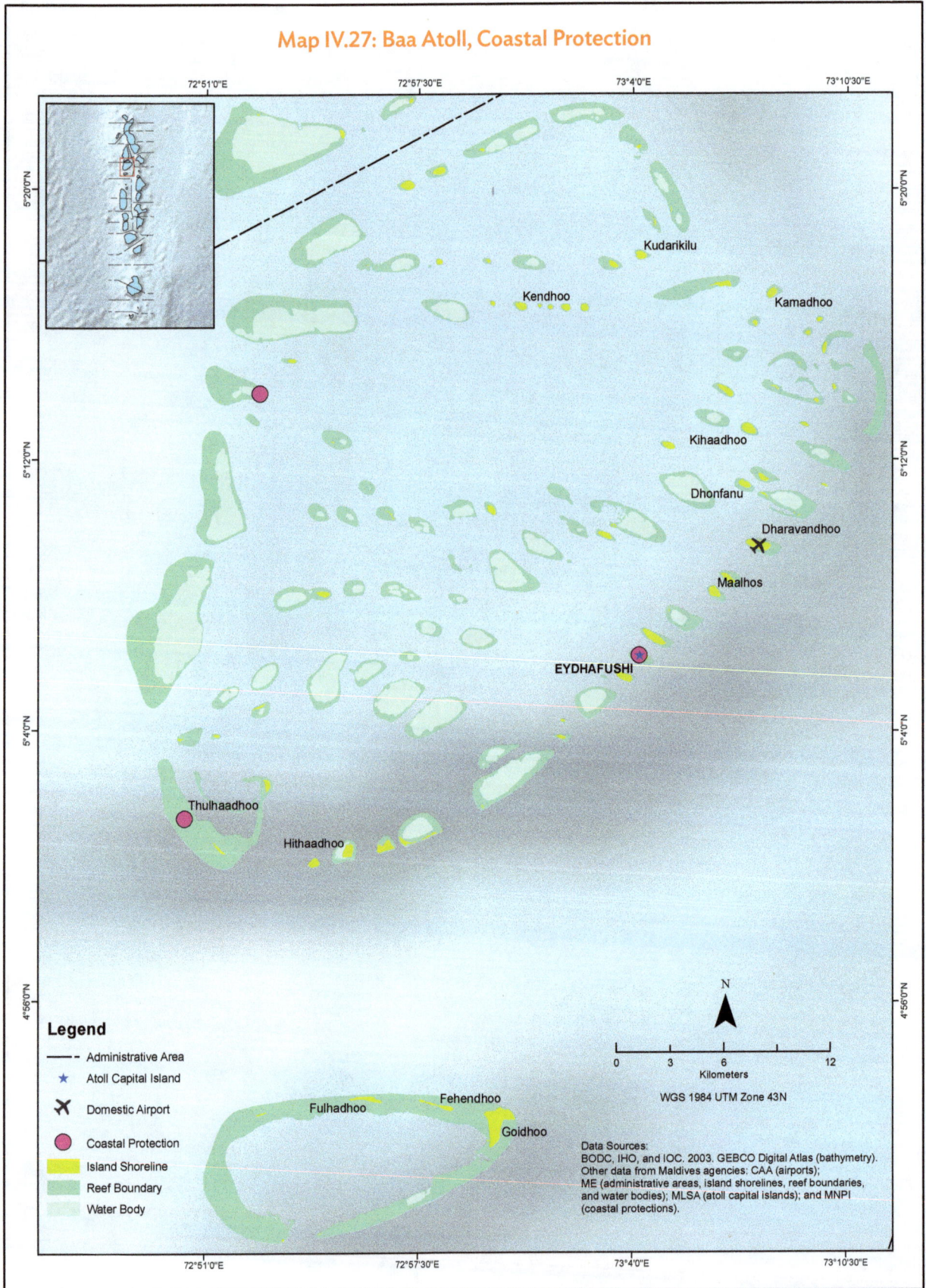

Kudarikilu

Kendhoo

Kamadhoo

Kihaadhoo

Dhonfanu

Dharavandhoo

Maalhos

EYDHAFUSHI

Thulhaadhoo

Hithaadhoo

Fulhadhoo

Fehendhoo

Goidhoo

Legend

— Administrative Area

★ Atoll Capital Island

✈ Domestic Airport

● Coastal Protection

■ Island Shoreline

■ Reef Boundary

■ Water Body

N

0   3   6   12
Kilometers

WGS 1984 UTM Zone 43N

Data Sources:
BODC, IHO, and IOC. 2003. GEBCO Digital Atlas (bathymetry).
Other data from Maldives agencies: CAA (airports);
ME (administrative areas, island shorelines, reef boundaries,
and water bodies); MLSA (atoll capital islands); and MNPI
(coastal protections).

Map IV.28: Dhaalu Atoll, Coastal Protection

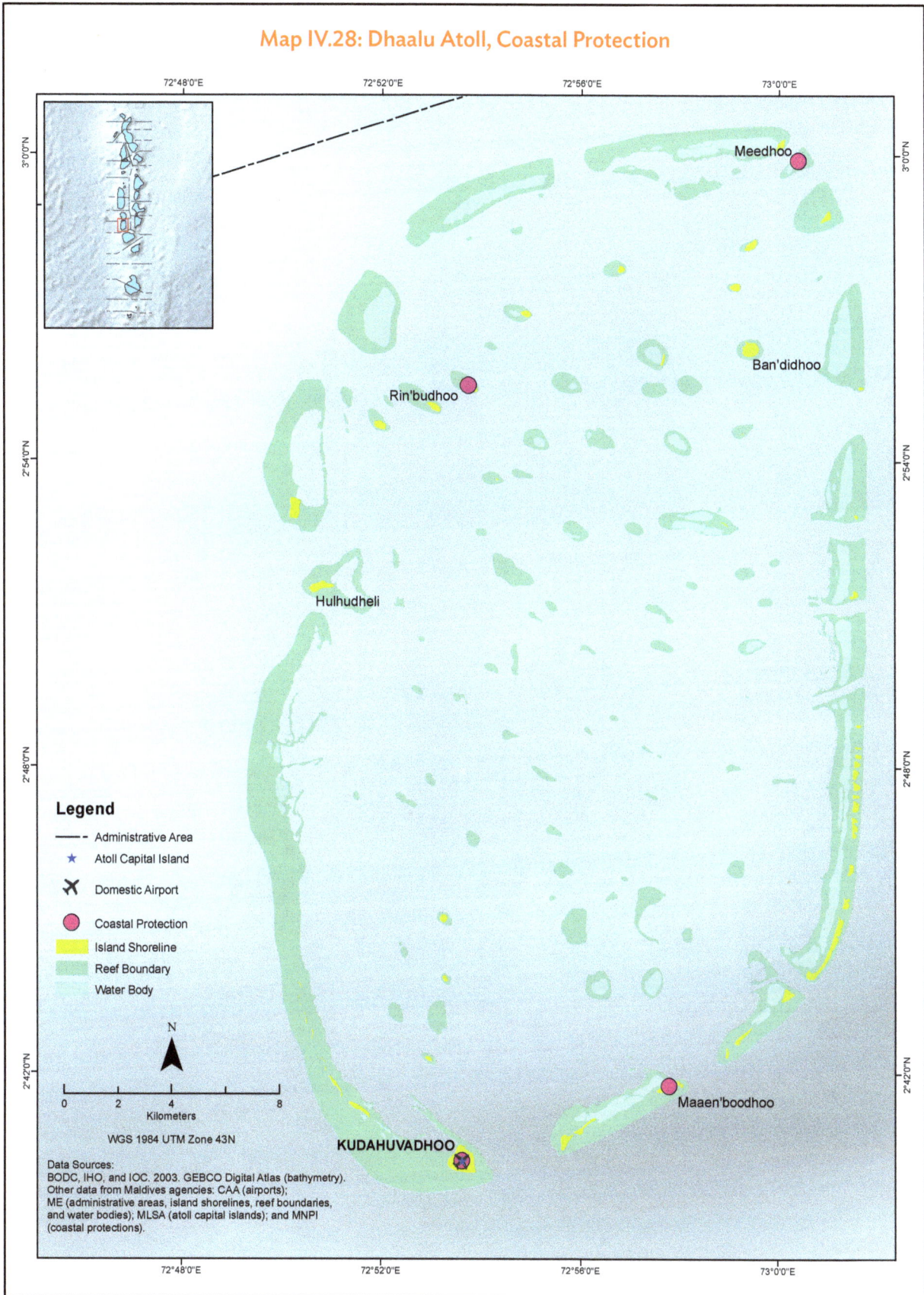

Meedhoo

Ban'didhoo

Rin'budhoo

Hulhudheli

Maaen'boodhoo

KUDAHUVADHOO

Legend

— · — Administrative Area

★ Atoll Capital Island

✈ Domestic Airport

⬤ Coastal Protection

▮ Island Shoreline

▮ Reef Boundary

▮ Water Body

N

0   2   4   8
Kilometers

WGS 1984 UTM Zone 43N

Data Sources:
BODC, IHO, and IOC. 2003. GEBCO Digital Atlas (bathymetry).
Other data from Maldives agencies: CAA (airports);
ME (administrative areas, island shorelines, reef boundaries,
and water bodies); MLSA (atoll capital islands); and MNPI
(coastal protections).

Map IV.29: Faafu Atoll, Coastal Protection

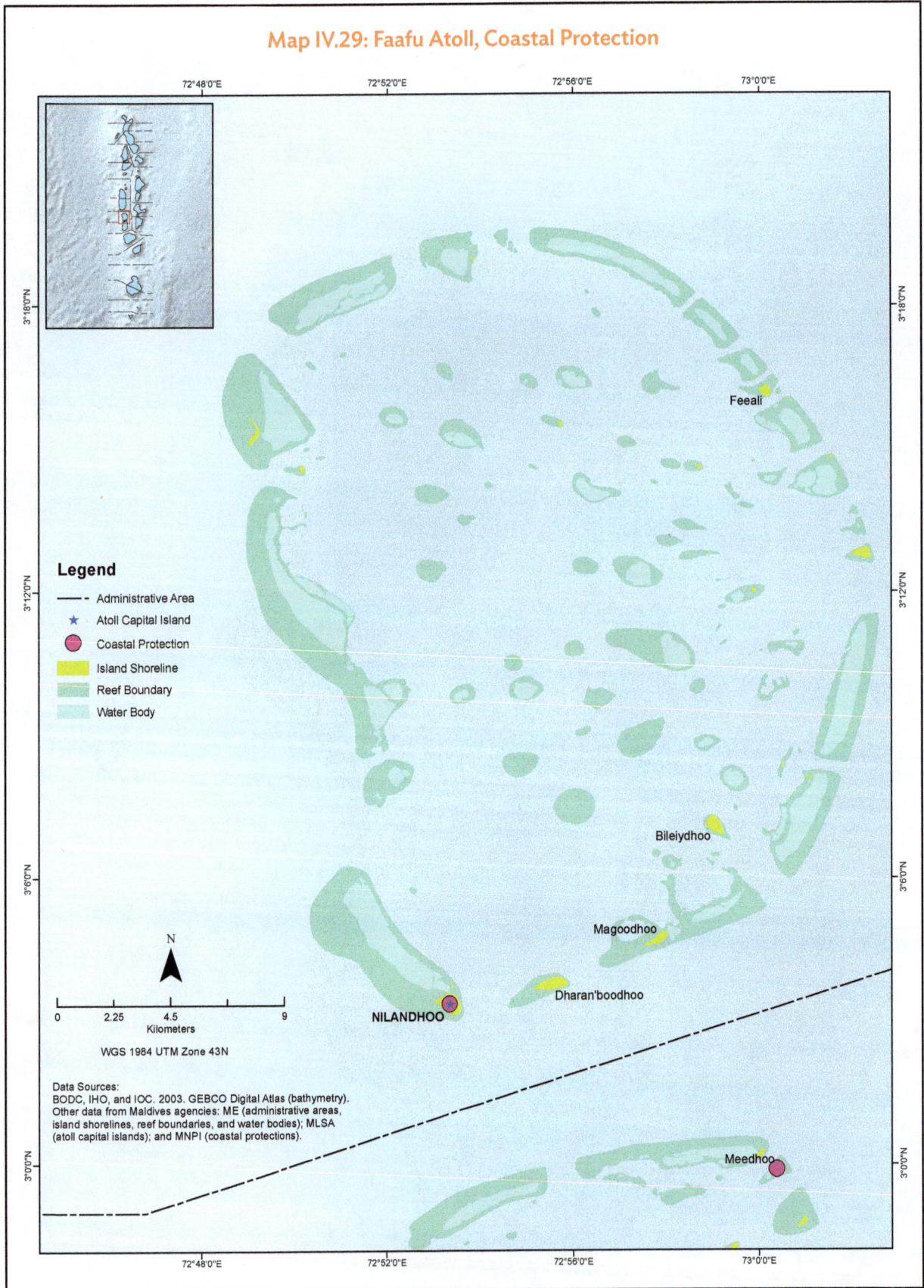

Feeali

Bileiydhoo

Magoodhoo

Dharan'boodhoo

NILANDHOO

Meedhoo

Legend

- - - Administrative Area
★ Atoll Capital Island
● Coastal Protection
Island Shoreline
Reef Boundary
Water Body

N

0   2.25   4.5   9
Kilometers

WGS 1984 UTM Zone 43N

Data Sources:
BODC, IHO, and IOC. 2003. GEBCO Digital Atlas (bathymetry).
Other data from Maldives agencies: ME (administrative areas,
island shorelines, reef boundaries, and water bodies); MLSA
(atoll capital islands); and MNPI (coastal protections).

Map IV.30: Gaafu Alifu Atoll, Coastal Protection

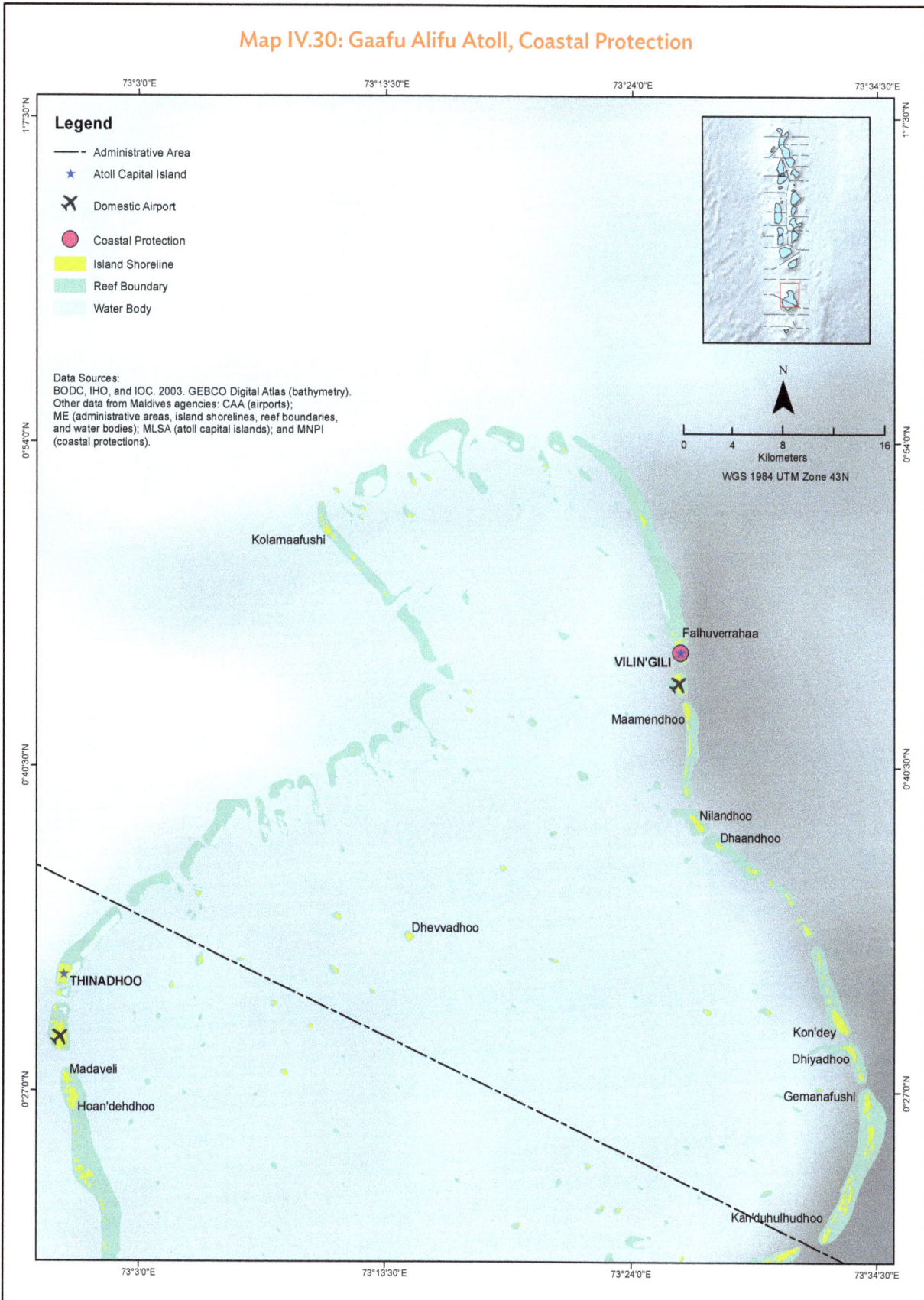

Legend

- – – Administrative Area
- ★ Atoll Capital Island
- ✈ Domestic Airport
- ⬤ Coastal Protection
- ▮ Island Shoreline
- ▮ Reef Boundary
- Water Body

Data Sources:
BODC, IHO, and IOC. 2003. GEBCO Digital Atlas (bathymetry).
Other data from Maldives agencies: CAA (airports);
ME (administrative areas, island shorelines, reef boundaries,
and water bodies); MLSA (atoll capital islands); and MNPI
(coastal protections).

N

0   4   8   16
Kilometers
WGS 1984 UTM Zone 43N

Kolamaafushi

Falhuverrahaa
VILIN'GILI ⬤
✈
Maamendhoo

Nilandhoo
Dhaandhoo

Dhevvadhoo

★ THINADHOO

✈

Madaveli

Hoan'dehdhoo

Kon'dey
Dhiyadhoo
Gemanafushi

Kanduhulhudhoo

Map IV.31: Gaafu Dhaalu Atoll, Coastal Protection

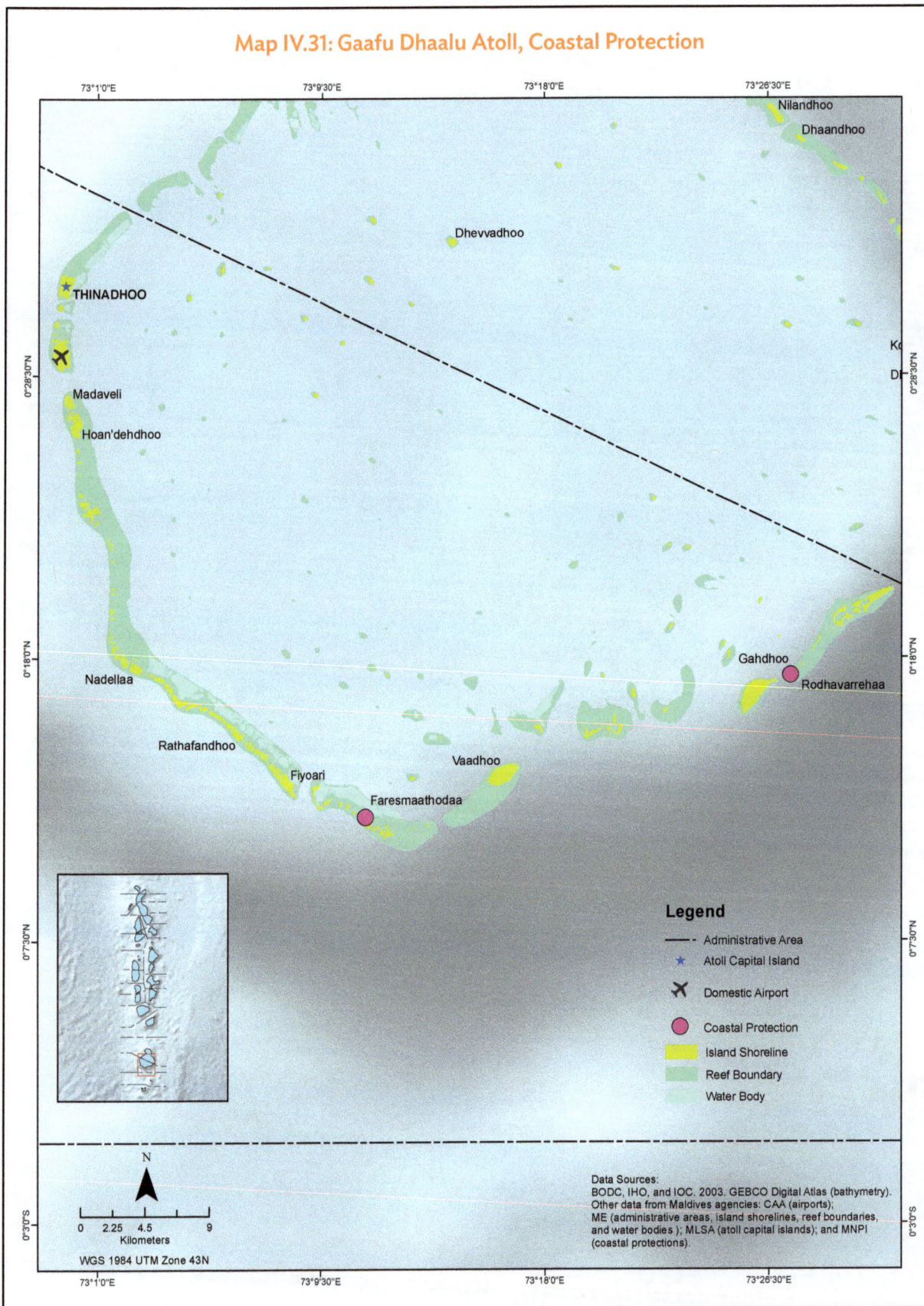

Nilandhoo

Dhaandhoo

Dhevvadhoo

THINADHOO

Madaveli

Hoan'dehdhoo

Gahdhoo

Nadellaa

Rodhavarrehaa

Rathafandhoo

Vaadhoo

Fiyoari

Faresmaathodaa

Legend

— Administrative Area

★ Atoll Capital Island

✈ Domestic Airport

⬤ Coastal Protection

▮ Island Shoreline

▮ Reef Boundary

▮ Water Body

N

0   2.25   4.5   9
Kilometers

WGS 1984 UTM Zone 43N

Data Sources:
BODC, IHO, and IOC, 2003. GEBCO Digital Atlas (bathymetry).
Other data from Maldives agencies: CAA (airports);
ME (administrative areas, island shorelines, reef boundaries,
and water bodies); MLSA (atoll capital islands); and MNPI
(coastal protections).

Map IV.32: Gnaviyani Atoll, Coastal Protection

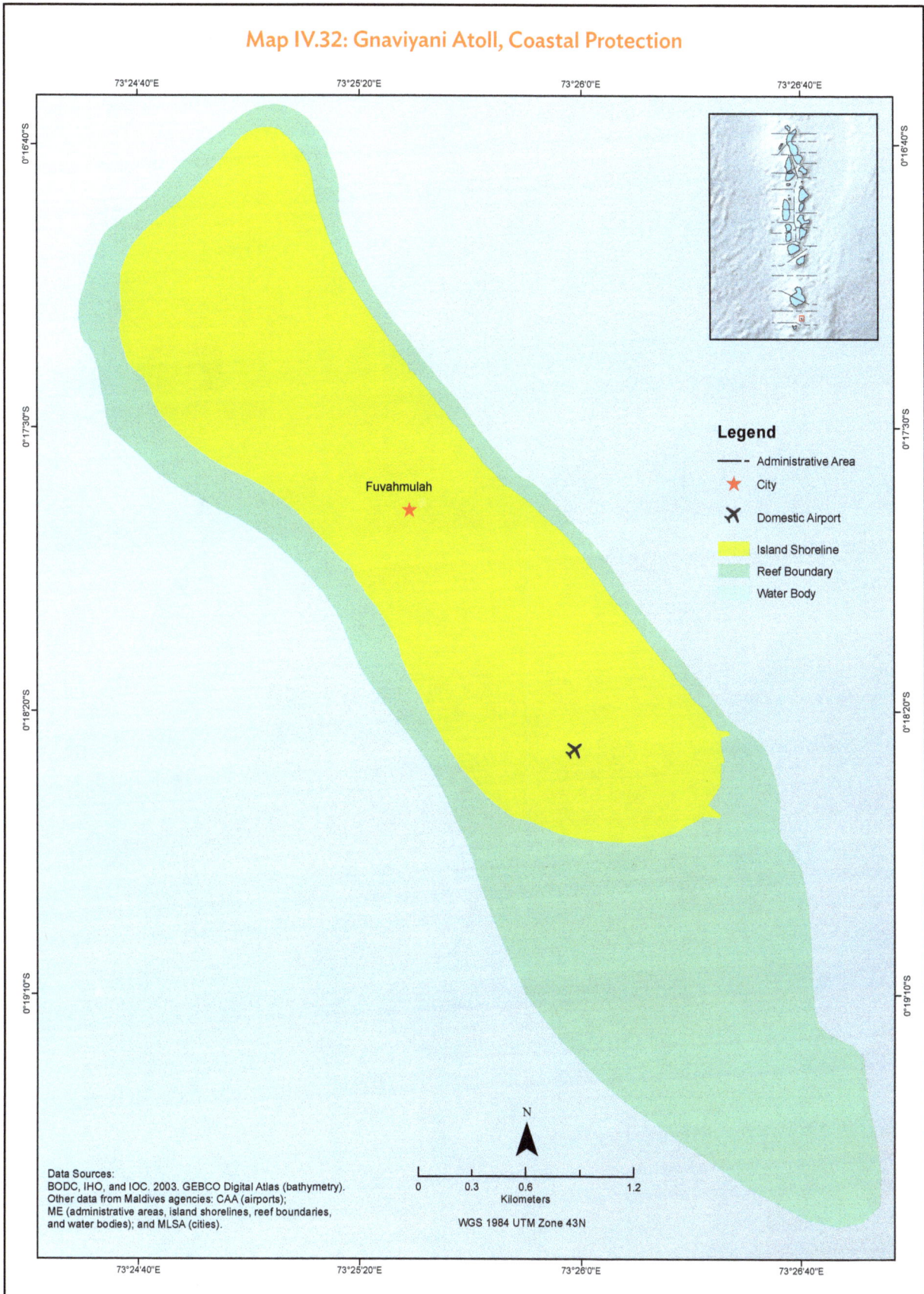

73°24'40"E   73°25'20"E   73°26'0"E   73°26'40"E

0°16'40"S

0°17'30"S

Fuvahmulah
★

✈

0°18'20"S

0°19'10"S

Legend

— — Administrative Area
★ City
✈ Domestic Airport
▇ Island Shoreline
▇ Reef Boundary
▇ Water Body

N

0   0.3   0.6   1.2
Kilometers

WGS 1984 UTM Zone 43N

Data Sources:
BODC, IHO, and IOC. 2003. GEBCO Digital Atlas (bathymetry).
Other data from Maldives agencies: CAA (airports);
ME (administrative areas, island shorelines, reef boundaries,
and water bodies); and MLSA (cities).

Map IV.33: Haa Alifu Atoll, Coastal Protection

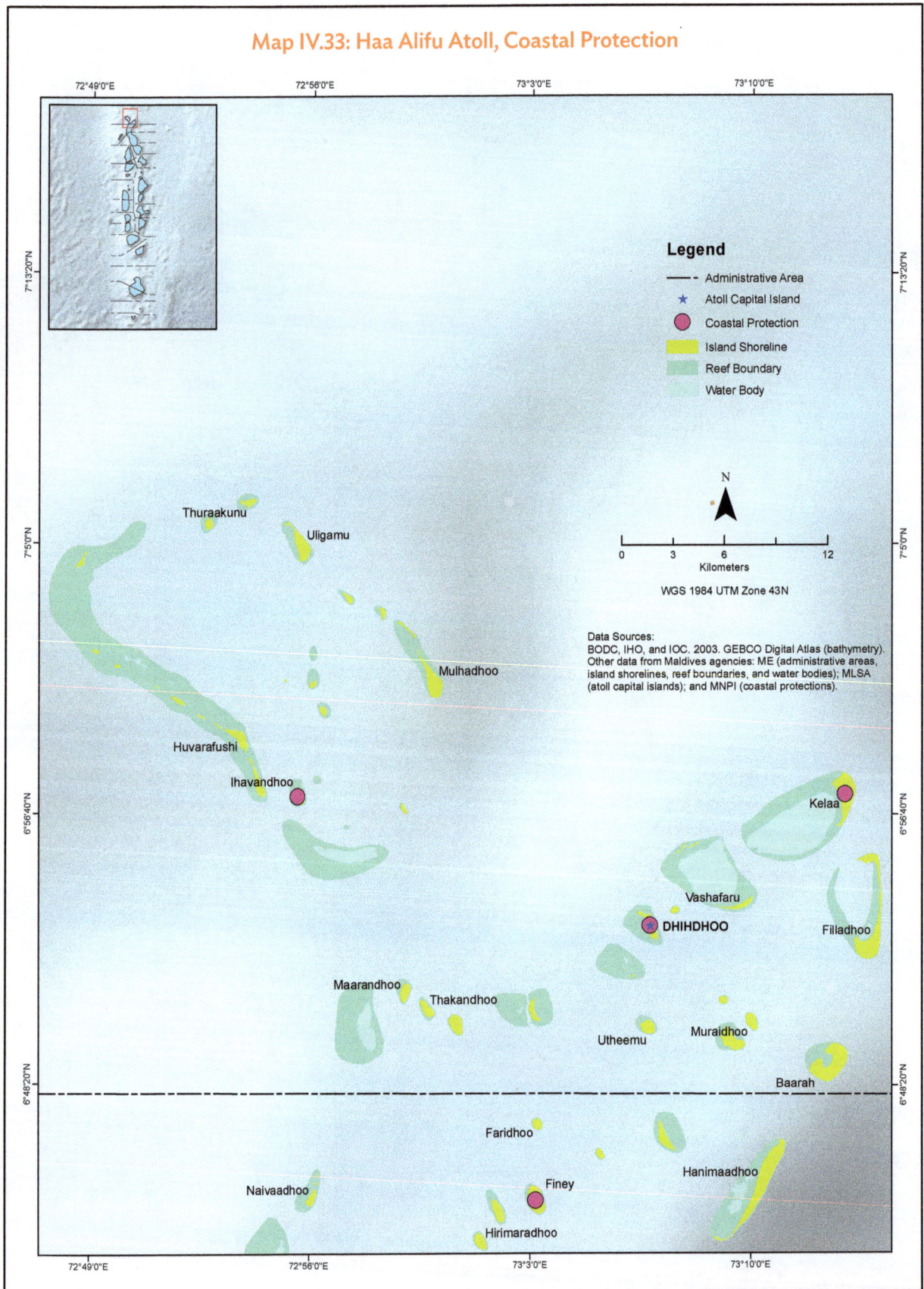

Legend

—— Administrative Area

★ Atoll Capital Island

● Coastal Protection

▮ Island Shoreline

▮ Reef Boundary

▮ Water Body

N

0   3   6   12
Kilometers

WGS 1984 UTM Zone 43N

Data Sources:
BODC, IHO, and IOC. 2003. GEBCO Digital Atlas (bathymetry).
Other data from Maldives agencies: ME (administrative areas,
island shorelines, reef boundaries, and water bodies); MLSA
(atoll capital islands); and MNPI (coastal protections).

Thuraakunu

Uligamu

Mulhadhoo

Huvarafushi

Ihavandhoo

Kelaa

Vashafaru

DHIHDHOO

Filladhoo

Maarandhoo

Thakandhoo

Muraidhoo

Utheemu

Baarah

Faridhoo

Hanimaadhoo

Naivaadhoo

Finey

Hirimaradhoo

Map IV.34: Haa Dhaalu Atoll, Coastal Protection

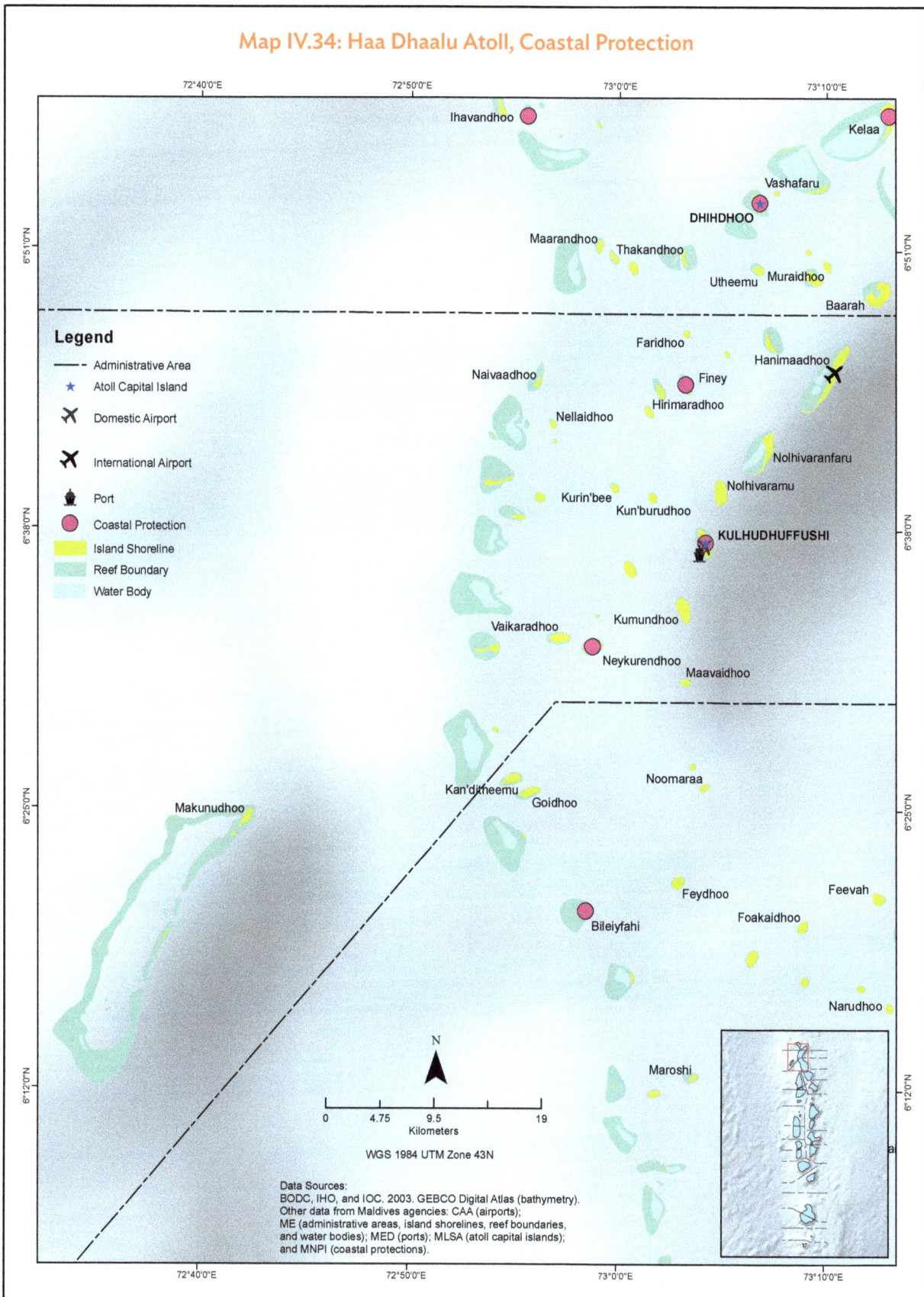

Ihavandhoo

Kelaa

Vashafaru

DHIHDHOO

Maarandhoo

Thakandhoo

Utheemu   Muraidhoo

Baarah

Faridhoo

Hanimaadhoo

Naivaadhoo

Finey

Hirimaradhoo

Nolhivaranfaru

Nellaidhoo

Nolhivaramu

Kurin'bee

Kun'burudhoo

KULHUDHUFFUSHI

Kumundhoo

Vaikaradhoo

Neykurendhoo

Maavaidhoo

Noomaraa

Kan'ditheemu

Makunudhoo

Goidhoo

Feydhoo   Feevah

Foakaidhoo

Bileiyfahi

Narudhoo

Maroshi

Legend

- Administrative Area
- ★ Atoll Capital Island
- ✈ Domestic Airport
- ✈ International Airport
- ⚓ Port
- ● Coastal Protection
- Island Shoreline
- Reef Boundary
- Water Body

N

```
0      4.75      9.5      19
            Kilometers
```

WGS 1984 UTM Zone 43N

Data Sources:
BODC, IHO, and IOC. 2003. GEBCO Digital Atlas (bathymetry).
Other data from Maldives agencies: CAA (airports);
ME (administrative areas, island shorelines, reef boundaries,
and water bodies); MED (ports); MLSA (atoll capital islands);
and MNPI (coastal protections).

72°40'0"E   72°50'0"E   73°0'0"E   73°10'0"E

6°51'0"N   6°38'0"N   6°25'0"N   6°12'0"N

Map IV.35: Laamu Atoll, Coastal Protection

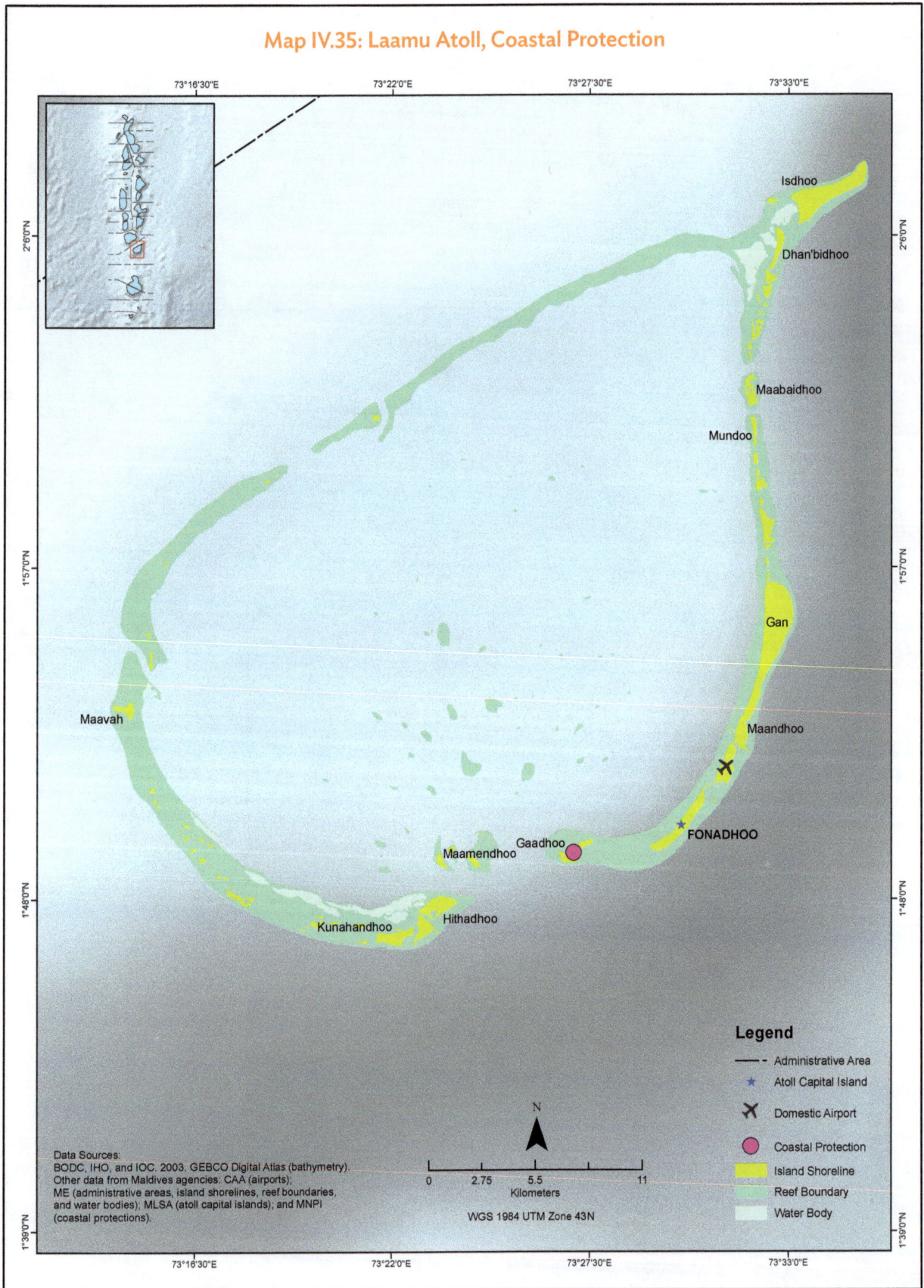

Isdhoo

Dhan'bidhoo

Maabaidhoo

Mundoo

Gan

Maavah

Maandhoo

FONADHOO

Gaadhoo

Maamendhoo

Kunahandhoo

Hithadhoo

Data Sources:
BODC, IHO, and IOC. 2003. GEBCO Digital Atlas (bathymetry).
Other data from Maldives agencies: CAA (airports);
ME (administrative areas, island shorelines, reef boundaries,
and water bodies); MLSA (atoll capital islands); and MNPI
(coastal protections).

N

0   2.75   5.5   11
Kilometers

WGS 1984 UTM Zone 43N

Legend

- – – Administrative Area
- ★ Atoll Capital Island
- ✕ Domestic Airport
- ⬤ Coastal Protection
- Island Shoreline
- Reef Boundary
- Water Body

Map IV.36: Lhaviyani Atoll, Coastal Protection

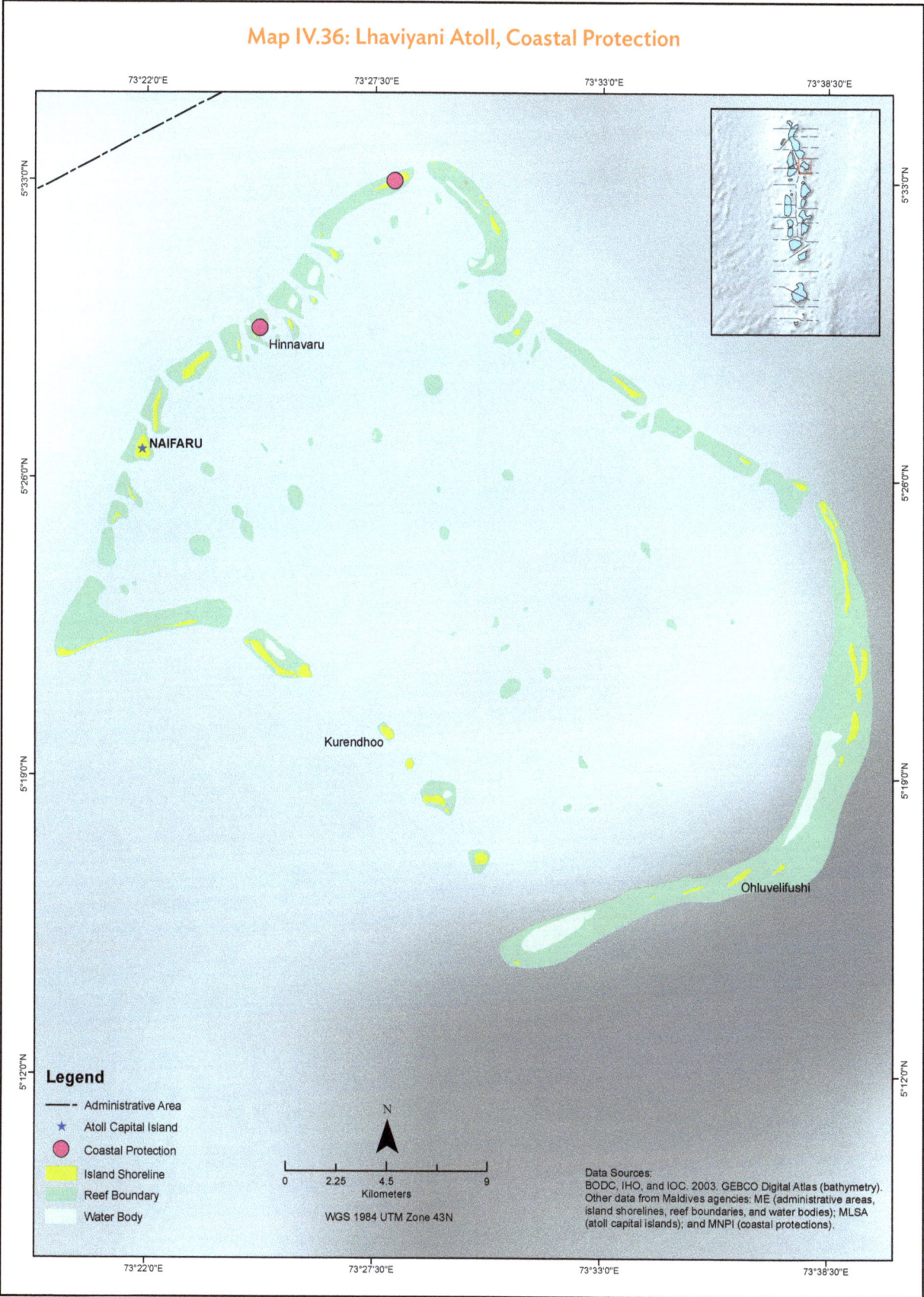

73°22'0"E   73°27'30"E   73°33'0"E   73°38'30"E

5°33'0"N

5°26'0"N

Hinnavaru

NAIFARU

5°19'0"N

Kurendhoo

Ohluvelifushi

5°12'0"N

Legend

– – – Administrative Area

★ Atoll Capital Island

● Coastal Protection

Island Shoreline

Reef Boundary

Water Body

N

0   2.25   4.5   9
Kilometers

WGS 1984 UTM Zone 43N

Data Sources:
BODC, IHO, and IOC. 2003. GEBCO Digital Atlas (bathymetry).
Other data from Maldives agencies: ME (administrative areas,
island shorelines, reef boundaries, and water bodies); MLSA
(atoll capital islands); and MNPI (coastal protections).

Map IV.37: Meemu Atoll, Coastal Protection

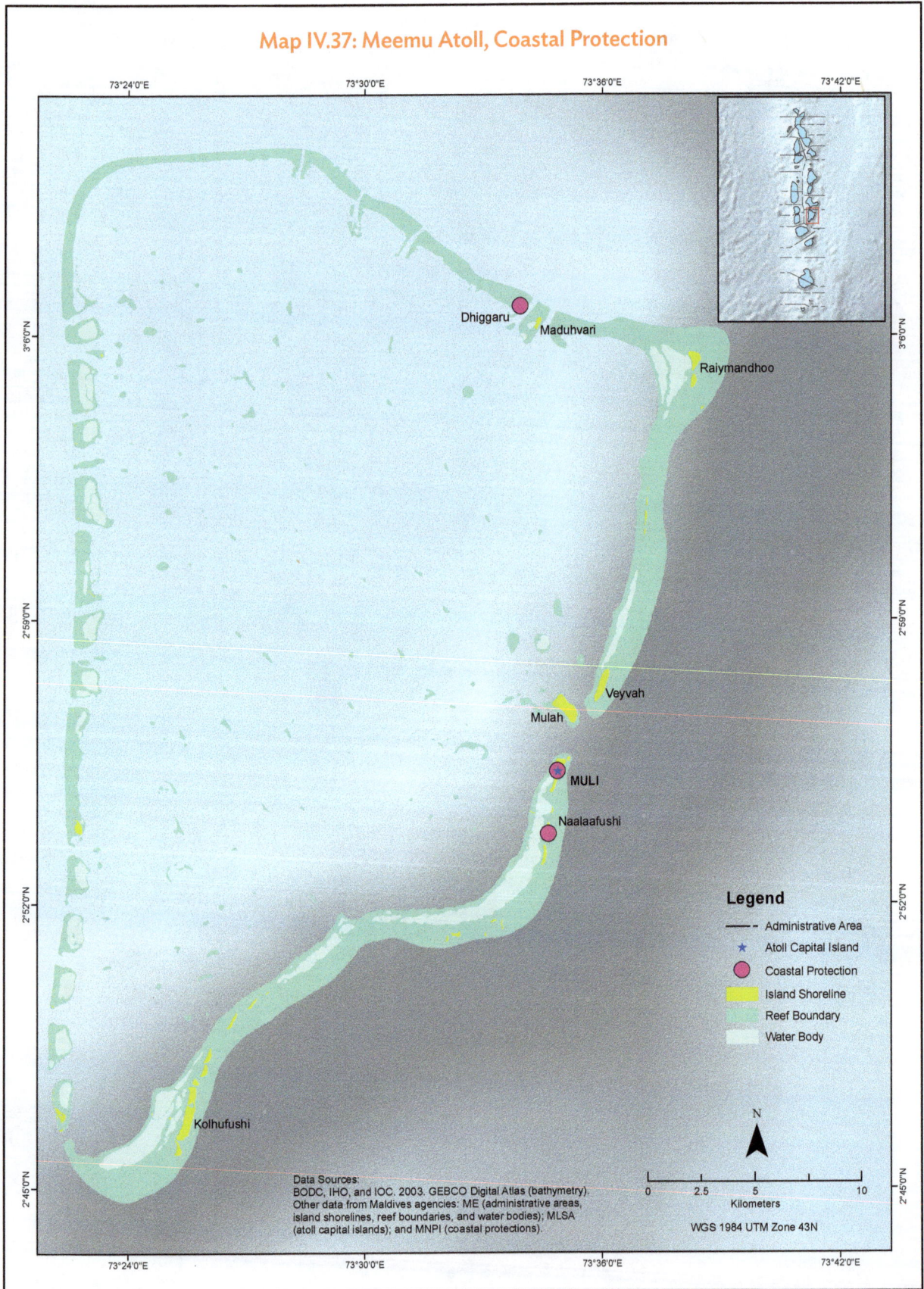

Dhiggaru

Maduhvari

Raiymandhoo

Veyvah

Mulah

MULI

Naalaafushi

Kolhufushi

Legend

— Administrative Area

★ Atoll Capital Island

● Coastal Protection

Island Shoreline

Reef Boundary

Water Body

Data Sources:
BODC, IHO, and IOC. 2003. GEBCO Digital Atlas (bathymetry).
Other data from Maldives agencies: ME (administrative areas,
island shorelines, reef boundaries, and water bodies); MLSA
(atoll capital islands); and MNPI (coastal protections).

0   2.5   5   10
Kilometers

WGS 1984 UTM Zone 43N

Map IV.38: Noonu Atoll, Coastal Protection

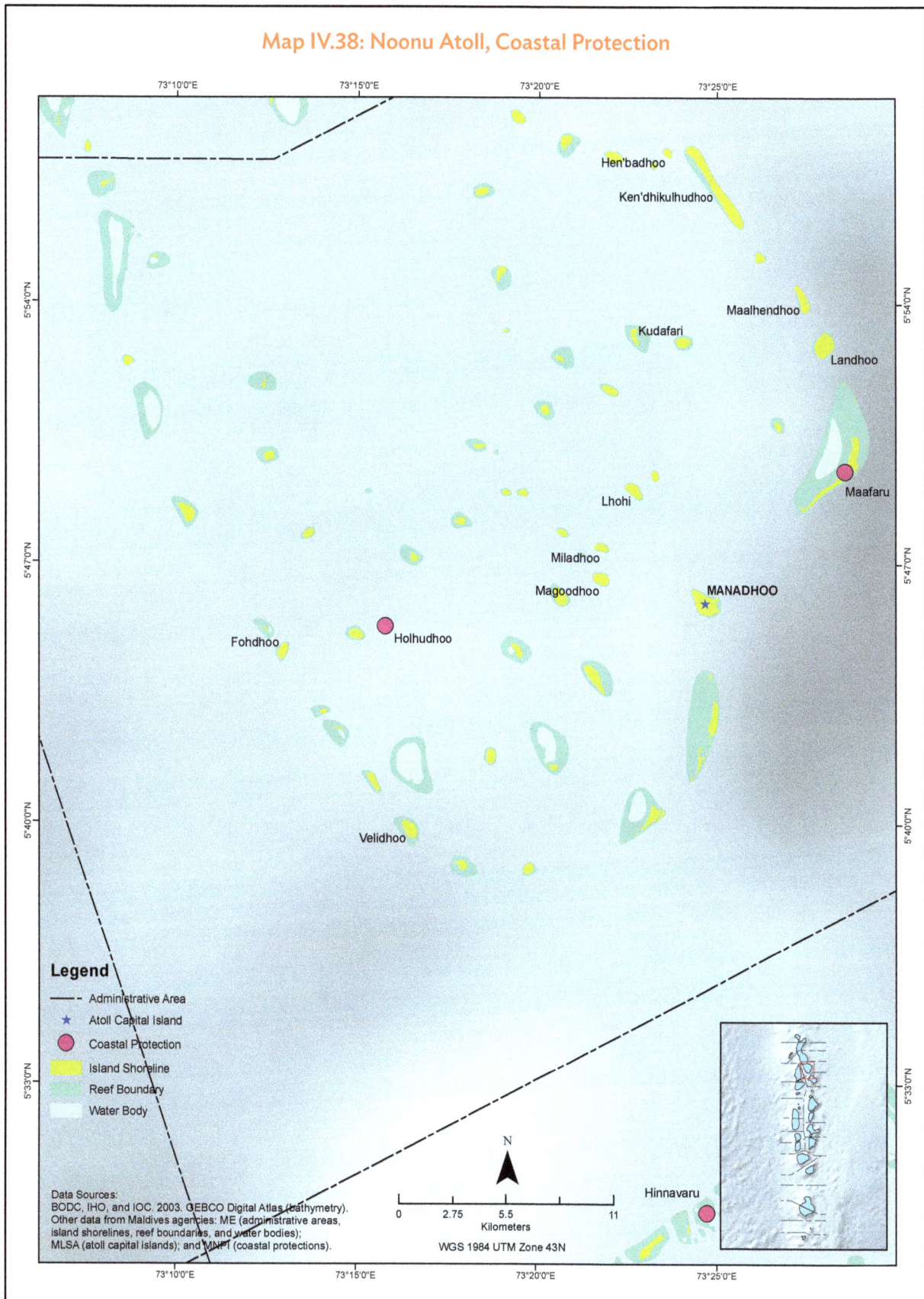

Hen'badhoo

Ken'dhikulhudhoo

Maalhendhoo

Kudafari

Landhoo

Maafaru

Lhohi

Miladhoo

MANADHOO

Magoodhoo

Fohdhoo

Holhudhoo

Velidhoo

Legend

- – – Administrative Area
- ★ Atoll Capital Island
- ● Coastal Protection
- ▢ Island Shoreline
- ▢ Reef Boundary
- ▢ Water Body

Data Sources:
BODC, IHO, and IOC. 2003. GEBCO Digital Atlas (bathymetry).
Other data from Maldives agencies: ME (administrative areas,
island shorelines, reef boundaries, and water bodies);
MLSA (atoll capital islands); and MNPI (coastal protections).

N

0   2.75   5.5   11
Kilometers

WGS 1984 UTM Zone 43N

Hinnavaru

Map IV.39: North Malé Atoll, Coastal Protection

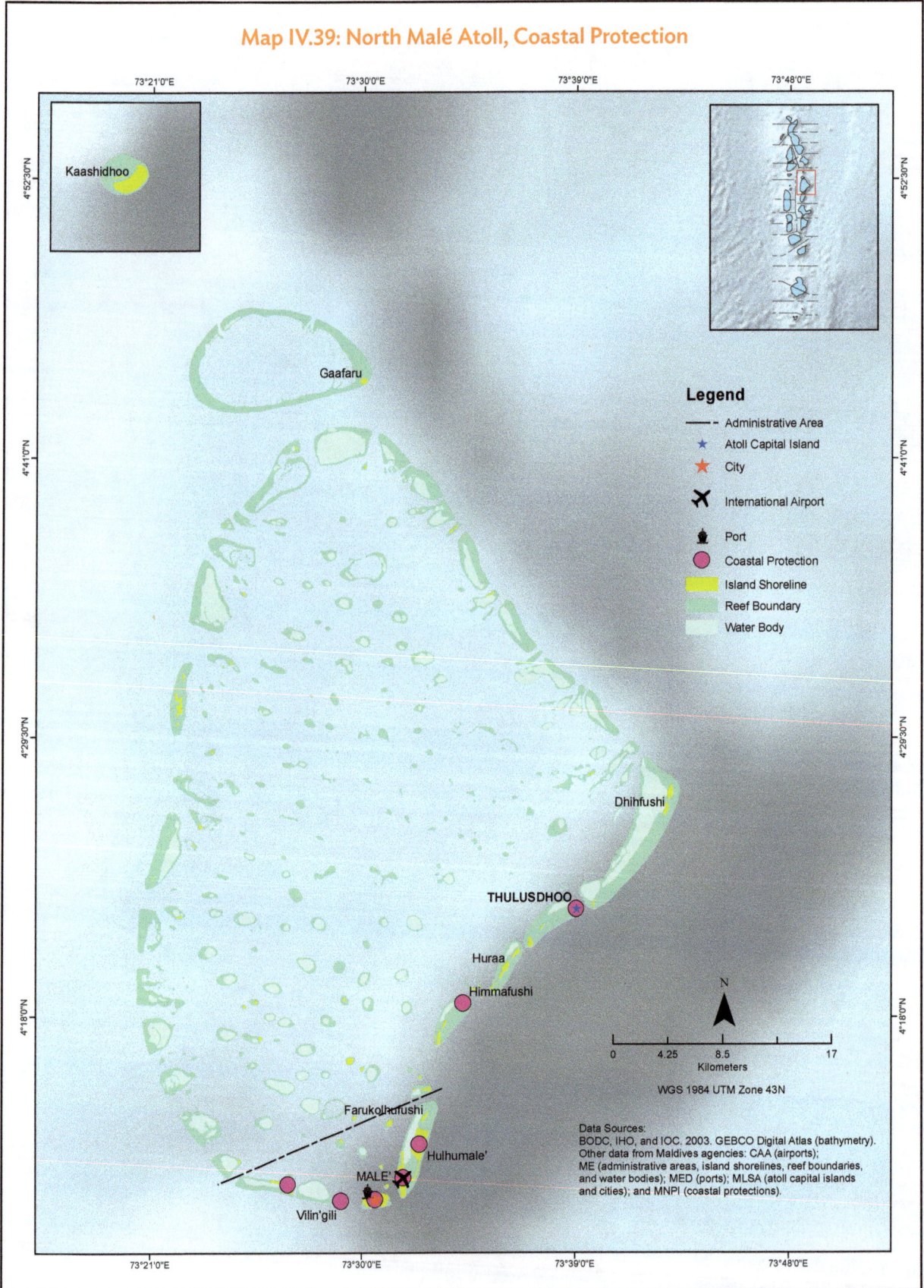

Kaashidhoo

Gaafaru

Legend

--- Administrative Area
★ Atoll Capital Island
★ City
✈ International Airport
⚓ Port
● Coastal Protection
◼ Island Shoreline
◼ Reef Boundary
◼ Water Body

Dhihfushi

THULUSDHOO

Huraa

Himmafushi

N

0   4.25   8.5   17
Kilometers

WGS 1984 UTM Zone 43N

Farukolhufushi

Hulhumale'

MALE' ✈

Vilin'gili

Data Sources:
BODC, IHO, and IOC. 2003. GEBCO Digital Atlas (bathymetry).
Other data from Maldives agencies: CAA (airports);
ME (administrative areas, island shorelines, reef boundaries,
and water bodies); MED (ports); MLSA (atoll capital islands
and cities); and MNPI (coastal protections).

Map IV.40: Raa Atoll, Coastal Protection

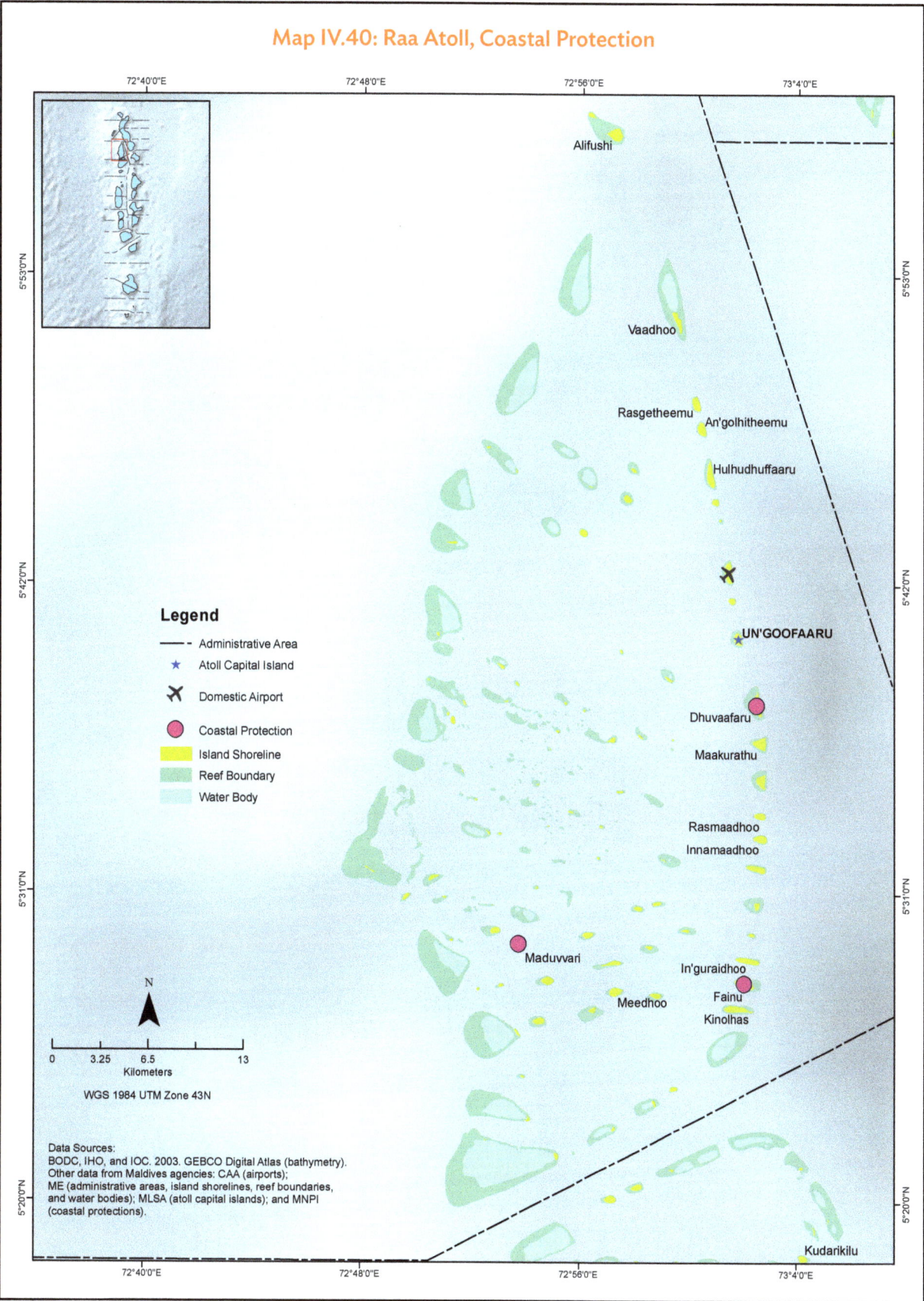

Alifushi

Vaadhoo

Rasgetheemu

An'golhitheemu

Hulhudhuffaaru

UN'GOOFAARU

Dhuvaafaru

Maakurathu

Rasmaadhoo

Innamaadhoo

In'guraidhoo

Maduvvari

Fainu

Meedhoo

Kinolhas

Kudarikilu

Legend

- – – – Administrative Area
- ★ Atoll Capital Island
- ✈ Domestic Airport
- ● Coastal Protection
- ▮ Island Shoreline
- ▮ Reef Boundary
- ▮ Water Body

N

0   3.25   6.5   13
Kilometers

WGS 1984 UTM Zone 43N

Data Sources:
BODC, IHO, and IOC. 2003. GEBCO Digital Atlas (bathymetry).
Other data from Maldives agencies: CAA (airports);
ME (administrative areas, island shorelines, reef boundaries,
and water bodies); MLSA (atoll capital islands); and MNPI
(coastal protections).

Map IV.41: Shaviyani Atoll, Coastal Protection

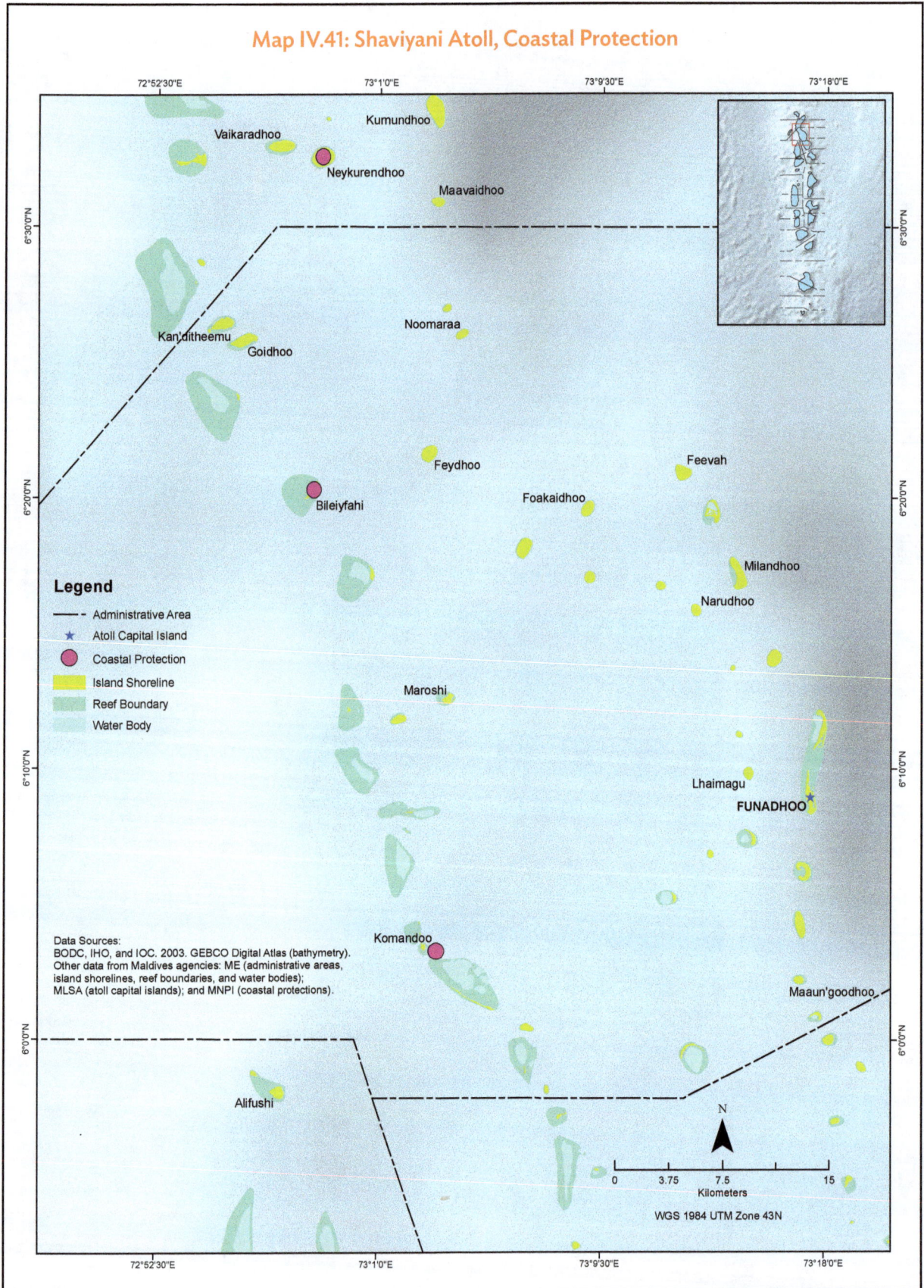

Vaikaradhoo

Kumundhoo

Neykurendhoo

Maavaidhoo

Kan'ditheemu

Goidhoo

Noomaraa

Feydhoo

Feevah

Bileiyfahi

Foakaidhoo

Milandhoo

Narudhoo

Legend

- - - Administrative Area
★ Atoll Capital Island
● Coastal Protection
▮ Island Shoreline
▮ Reef Boundary
▮ Water Body

Maroshi

Lhaimagu

FUNADHOO

Komandoo

Maaun'goodhoo

Data Sources:
BODC, IHO, and IOC. 2003. GEBCO Digital Atlas (bathymetry).
Other data from Maldives agencies: ME (administrative areas,
island shorelines, reef boundaries, and water bodies);
MLSA (atoll capital islands); and MNPI (coastal protections).

Alifushi

N

0   3.75   7.5   15
Kilometers

WGS 1984 UTM Zone 43N

Map IV.42: South Malé Atoll, Coastal Protection

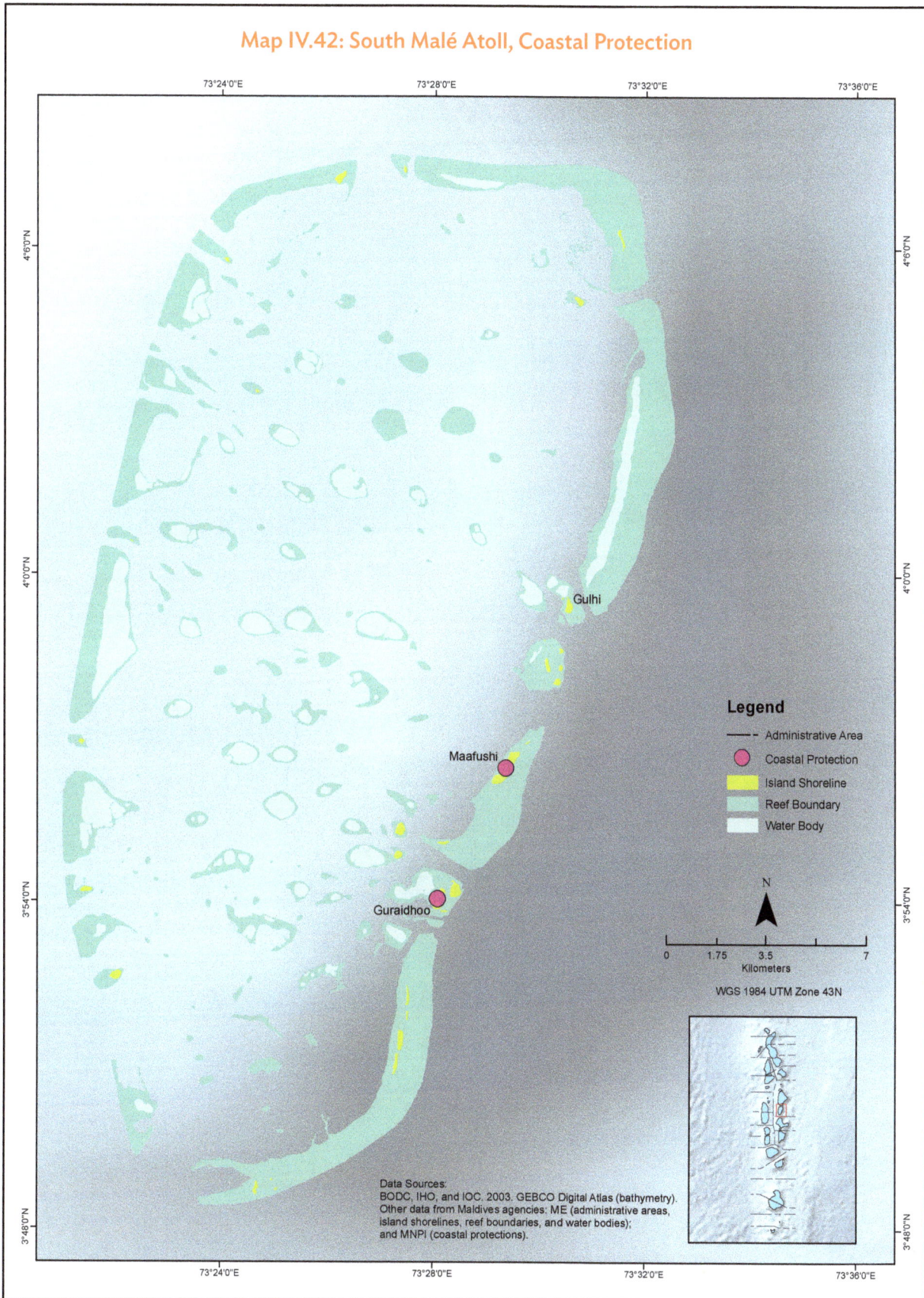

Legend

--- Administrative Area

● Coastal Protection

Island Shoreline

Reef Boundary

Water Body

N

0   1.75   3.5   7
Kilometers

WGS 1984 UTM Zone 43N

Gulhi

Maafushi

Guraidhoo

Data Sources:
BODC, IHO, and IOC. 2003. GEBCO Digital Atlas (bathymetry).
Other data from Maldives agencies: ME (administrative areas,
island shorelines, reef boundaries, and water bodies);
and MNPI (coastal protections).

Map IV.43: Thaa Atoll, Coastal Protection

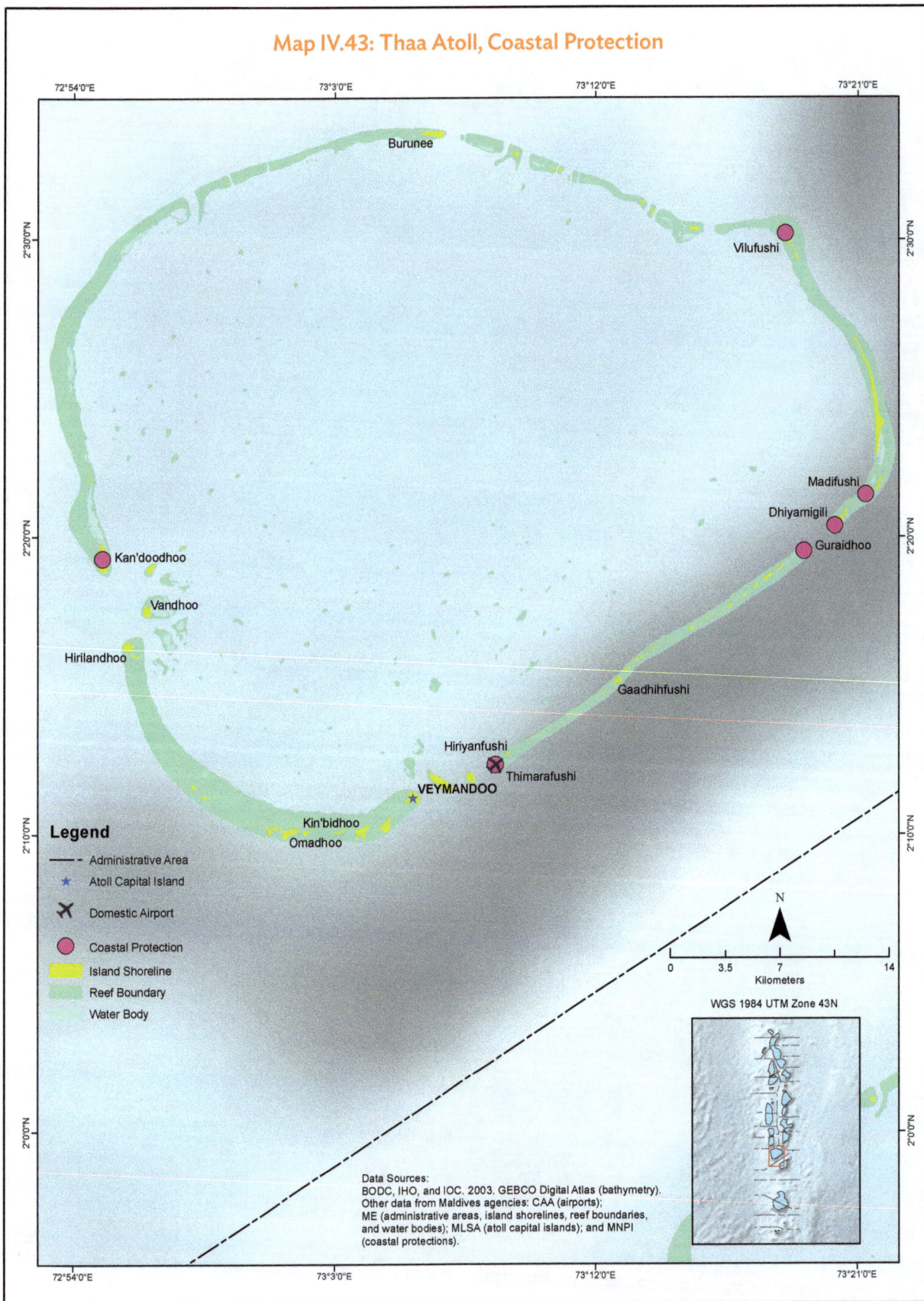

Burunee

Vilufushi

Madifushi

Dhiyamigili

Guraidhoo

Kan'doodhoo

Vandhoo

Hirilandhoo

Gaadhihfushi

Hiriyanfushi

Thimarafushi

VEYMANDOO

Kin'bidhoo

Omadhoo

Legend

- —— Administrative Area
- ★ Atoll Capital Island
- ✈ Domestic Airport
- ● Coastal Protection
- ▮ Island Shoreline
- ▮ Reef Boundary
- ▮ Water Body

N

0   3.5   7   14
Kilometers

WGS 1984 UTM Zone 43N

Data Sources:
BODC, IHO, and IOC. 2003. GEBCO Digital Atlas (bathymetry).
Other data from Maldives agencies: CAA (airports);
ME (administrative areas, island shorelines, reef boundaries,
and water bodies); MLSA (atoll capital islands); and MNPI
(coastal protections).

Map IV.44: Vaavu Atoll, Coastal Protection

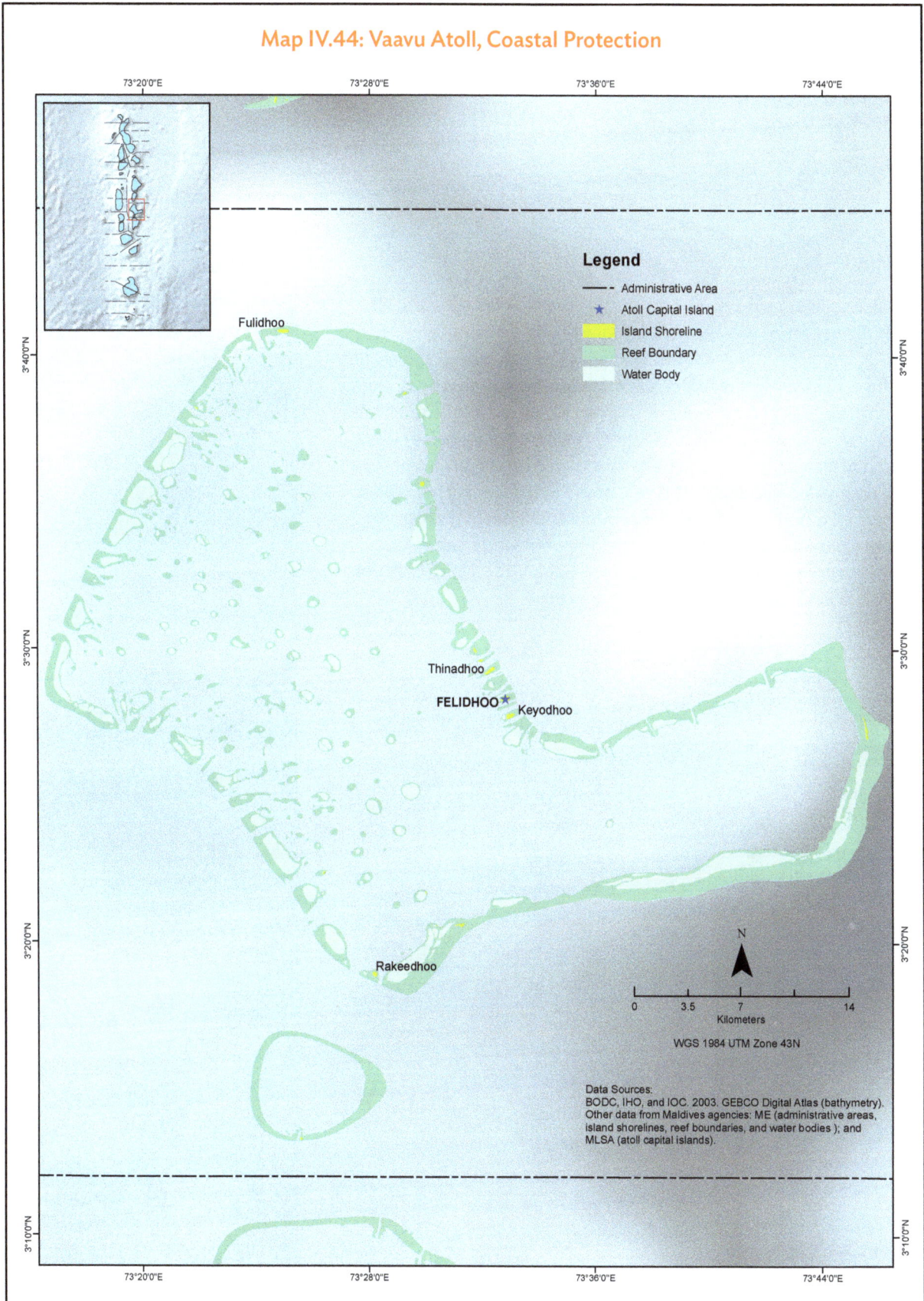

73°20'0"E   73°28'0"E   73°36'0"E   73°44'0"E

Legend

- - - Administrative Area
★ Atoll Capital Island
▮ Island Shoreline
▮ Reef Boundary
▯ Water Body

Fulidhoo

Thinadhoo

FELIDHOO ★ Keyodhoo

Rakeedhoo

3°40'0"N

3°30'0"N

3°20'0"N

3°10'0"N

N

0   3.5   7   14
Kilometers

WGS 1984 UTM Zone 43N

Data Sources:
BODC, IHO, and IOC. 2003. GEBCO Digital Atlas (bathymetry).
Other data from Maldives agencies: ME (administrative areas,
island shorelines, reef boundaries, and water bodies); and
MLSA (atoll capital islands).

Marine Conservation and Biodiversity

Maldives has greater coral reef coverage than available dry land. It has diverse and rich coral reefs, lagoons teeming with aquatic life, mangrove ecosystems where fingerlings as well as avian creatures find refuge, beaches where turtles lay their eggs, and seagrass beds. Human activities, coastal erosion, and changing climate patterns—specifically warmer seas—threaten the rich marine resources, particularly leading to coral bleaching. To protect this, the Marine Research Centre (now called 'Marine Research Institute') established coral reef monitoring sites in the 1990s to address the coral bleaching event that occurred in 1998 (Ibrahim et al. 2017). Coral reef monitoring sites can be found in the islands of Addu City and Alifu Alifu, Gaafu Alifu, Haa Dhaalu, North Malé, Seemu, and Vaavu atolls.

Table IV.2: Maldives, Islands with Coral Reef Monitoring Sites

Atoll	Island
Haa Dhaalu	Hondaafushi
Haa Dhaalu	Finey
Haa Dhaalu	Hirimaradhoo
North Malé	Bodubandos
North Malé	Udhafushi
North Malé	Enboodhoofinolhu
Alifu Alifu	Fesdhoo
Alifu Alifu	Maayaafushi
Alifu Alifu	Velidhoo
Alifu Alifu	Kandhonlhudhoo
Vaavu	Anbaraa
Vaavu	Vattaru
Vaavu	Foththeyo
Gaafu Alifu	Kooddoo
Seemu	Hithadhoo
Seemu	Gan
Seemu	Vilingili

Source: Maldives Marine Research Centre, 2017.

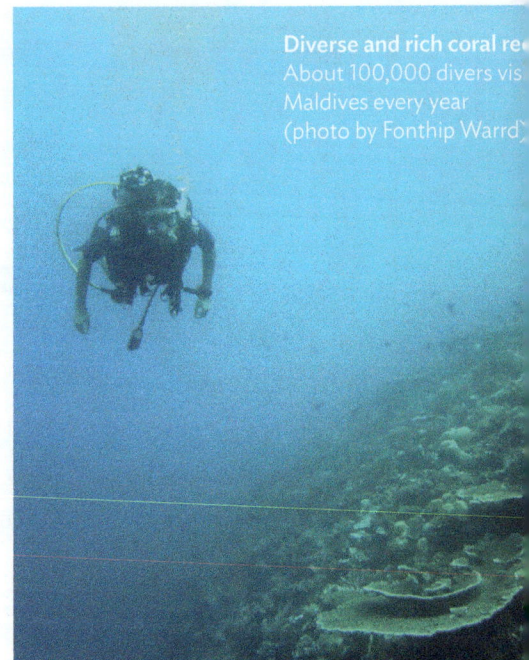

Diverse and rich coral ree[f]
About 100,000 divers vis[it]
Maldives every year
(photo by Fonthip Warrd)

Aquatic life. Moray eels
on corals (photo by Julian
Svoboda)

Biodiversity. Coral reef teeming with life (photo by Fonthip Warrd).

Map IV.45: Addu City, Biodiversity

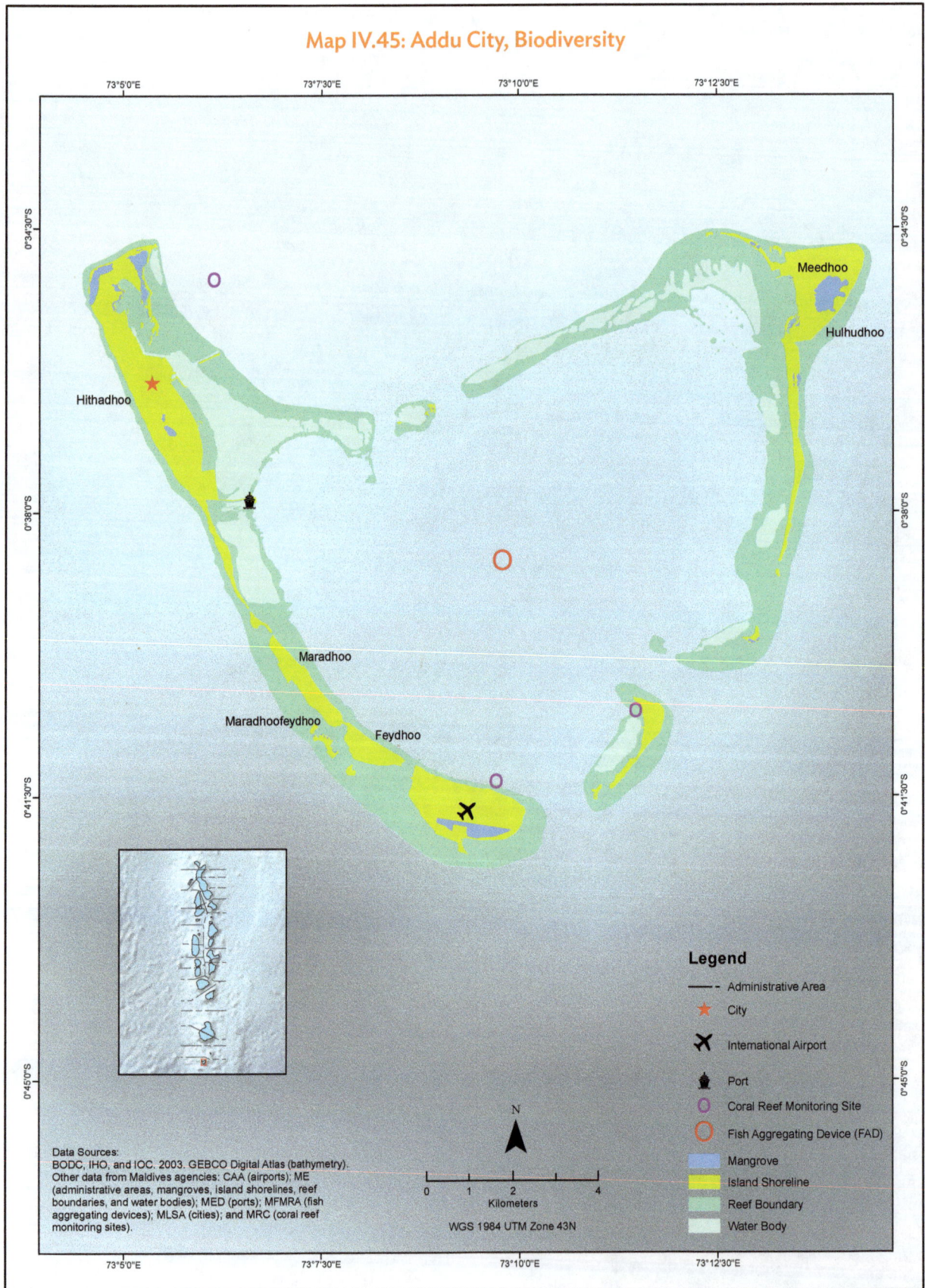

Meedhoo

Hulhudhoo

Hithadhoo

Maradhoo

Maradhoofeydhoo

Feydhoo

Legend

- — - Administrative Area
- ★ City
- ✈ International Airport
- ⚓ Port
- ○ Coral Reef Monitoring Site
- ○ Fish Aggregating Device (FAD)
- ▉ Mangrove
- ▉ Island Shoreline
- ▉ Reef Boundary
- ▉ Water Body

Data Sources:
BODC, IHO, and IOC. 2003. GEBCO Digital Atlas (bathymetry).
Other data from Maldives agencies: CAA (airports); ME
(administrative areas, mangroves, island shorelines, reef
boundaries, and water bodies); MED (ports); MFMRA (fish
aggregating devices); MLSA (cities); and MRC (coral reef
monitoring sites).

N

0   1   2   4
Kilometers

WGS 1984 UTM Zone 43N

Map IV.46: Alifu Alifu Atoll, Biodiversity

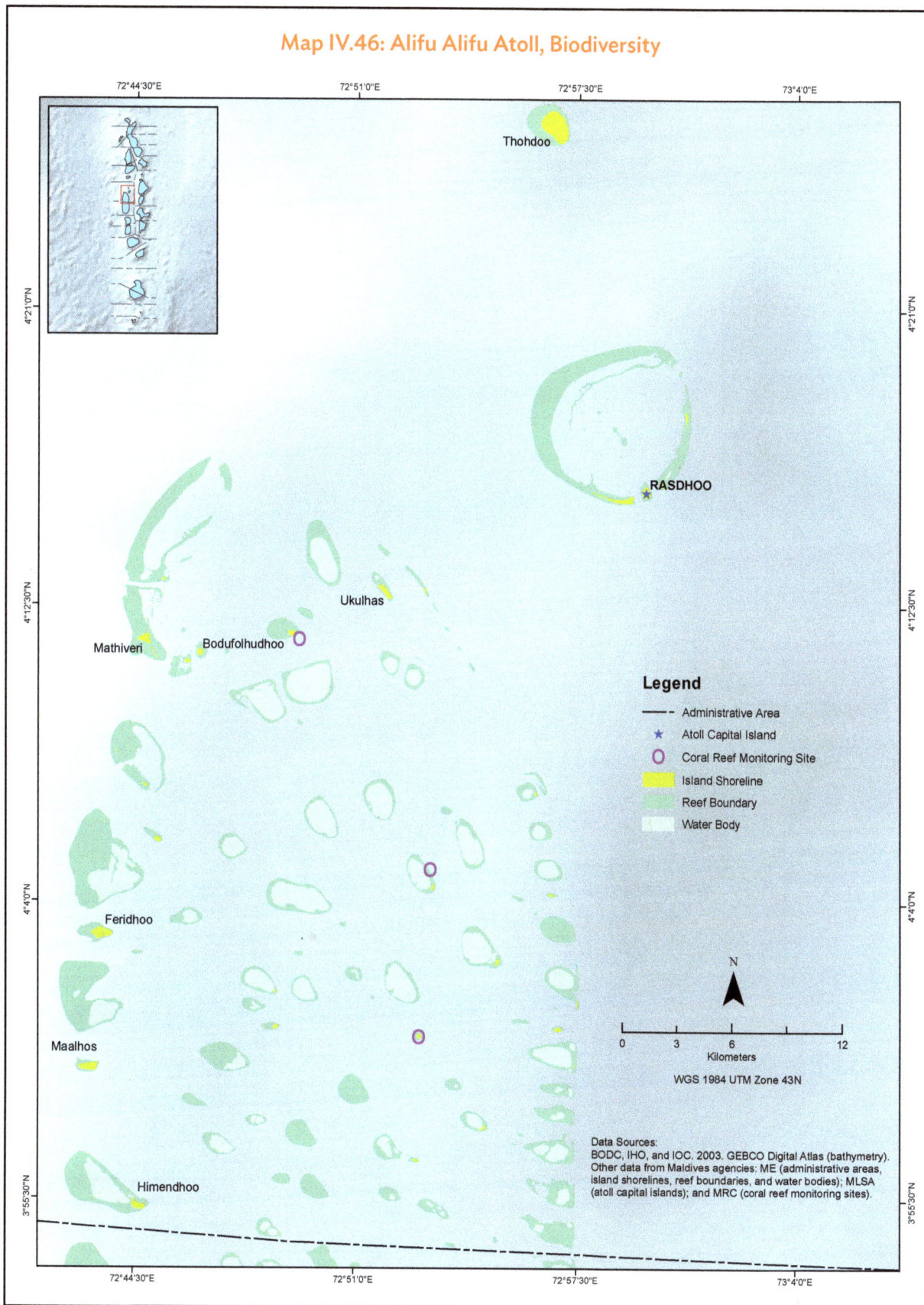

Thohdoo

RASDHOO

Ukulhas

Mathiveri   Bodufolhudhoo

Feridhoo

Maalhos

Himendhoo

Legend

- — — Administrative Area
- ★ Atoll Capital Island
- ⬡ Coral Reef Monitoring Site
- ▮ Island Shoreline
- ▮ Reef Boundary
- ▮ Water Body

N

| 0 | 3 | 6 | 12 |

Kilometers

WGS 1984 UTM Zone 43N

Data Sources:
BODC, IHO, and IOC. 2003. GEBCO Digital Atlas (bathymetry).
Other data from Maldives agencies: ME (administrative areas,
island shorelines, reef boundaries, and water bodies); MLSA
(atoll capital islands); and MRC (coral reef monitoring sites).

Map IV.47: Alifu Dhaalu Atoll, Biodiversity

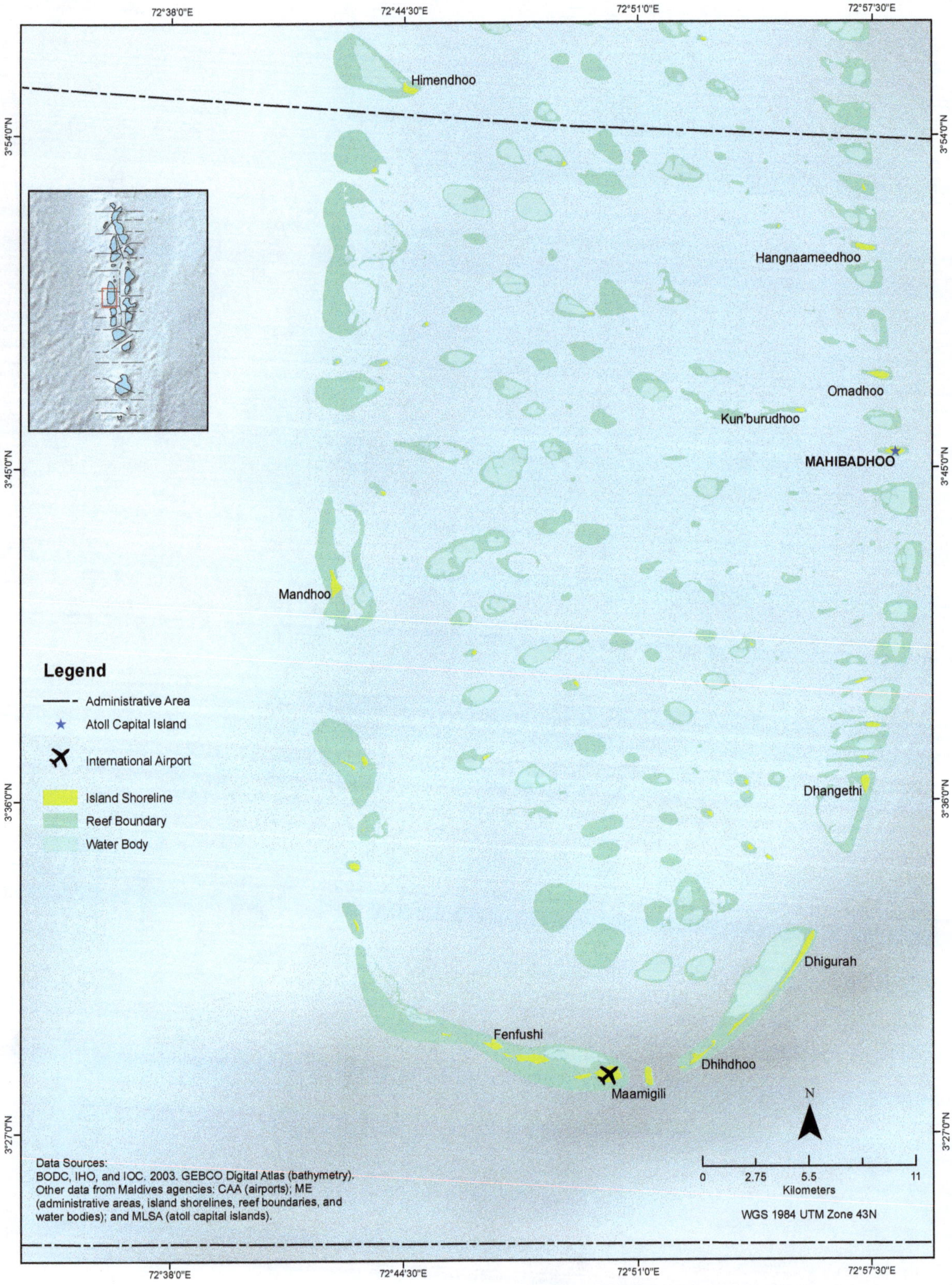
72°38'0"E
72°44'30"E
72°51'0"E
72°57'30"E
3°54'0"N
3°45'0"N
3°36'0"N
3°27'0"N
Himendhoo
Hangnaameedhoo
Omadhoo
Kun'burudhoo
MAHIBADHOO
Mandhoo
Dhangethi
Dhigurah
Fenfushi
Dhihdhoo
Maamigili
Legend
Administrative Area
Atoll Capital Island
International Airport
Island Shoreline
Reef Boundary
Water Body
N
0
2.75
5.5
11
Kilometers
WGS 1984 UTM Zone 43N
Data Sources:
BODC, IHO, and IOC. 2003. GEBCO Digital Atlas (bathymetry).
Other data from Maldives agencies: CAA (airports); ME (administrative areas, island shorelines, reef boundaries, and water bodies); and MLSA (atoll capital islands).

Map IV.48: Baa Atoll, Biodiversity

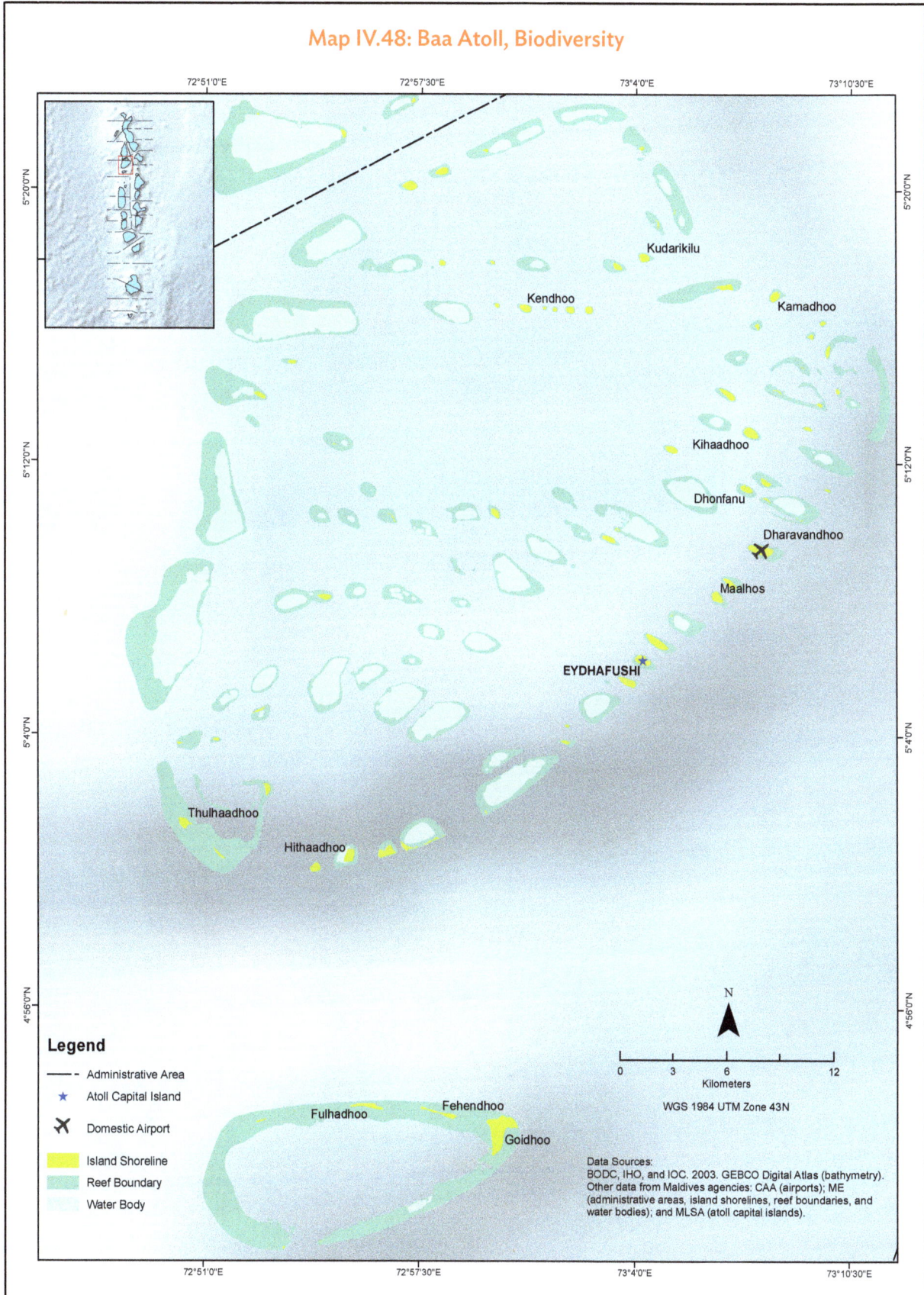

72°51'0"E   72°57'30"E   73°4'0"E   73°10'30"E

5°20'N

Kudarikilu

Kendhoo

Kamadhoo

Kihaadhoo

Dhonfanu

Dharavandhoo

Maalhos

EYDHAFUSHI

5°12'N

5°4'N

Thulhaadhoo

Hithaadhoo

4°56'N

N

Fulhadhoo   Fehendhoo

Goidhoo

Legend

- – – – Administrative Area
- ★ Atoll Capital Island
- ✈ Domestic Airport
- ▮ (yellow) Island Shoreline
- ▮ (green) Reef Boundary
- ▮ Water Body

0   3   6   12
Kilometers

WGS 1984 UTM Zone 43N

Data Sources:
BODC, IHO, and IOC. 2003. GEBCO Digital Atlas (bathymetry).
Other data from Maldives agencies: CAA (airports); ME
(administrative areas, island shorelines, reef boundaries, and
water bodies); and MLSA (atoll capital islands).

Map IV.49: Dhaalu Atoll, Biodiversity

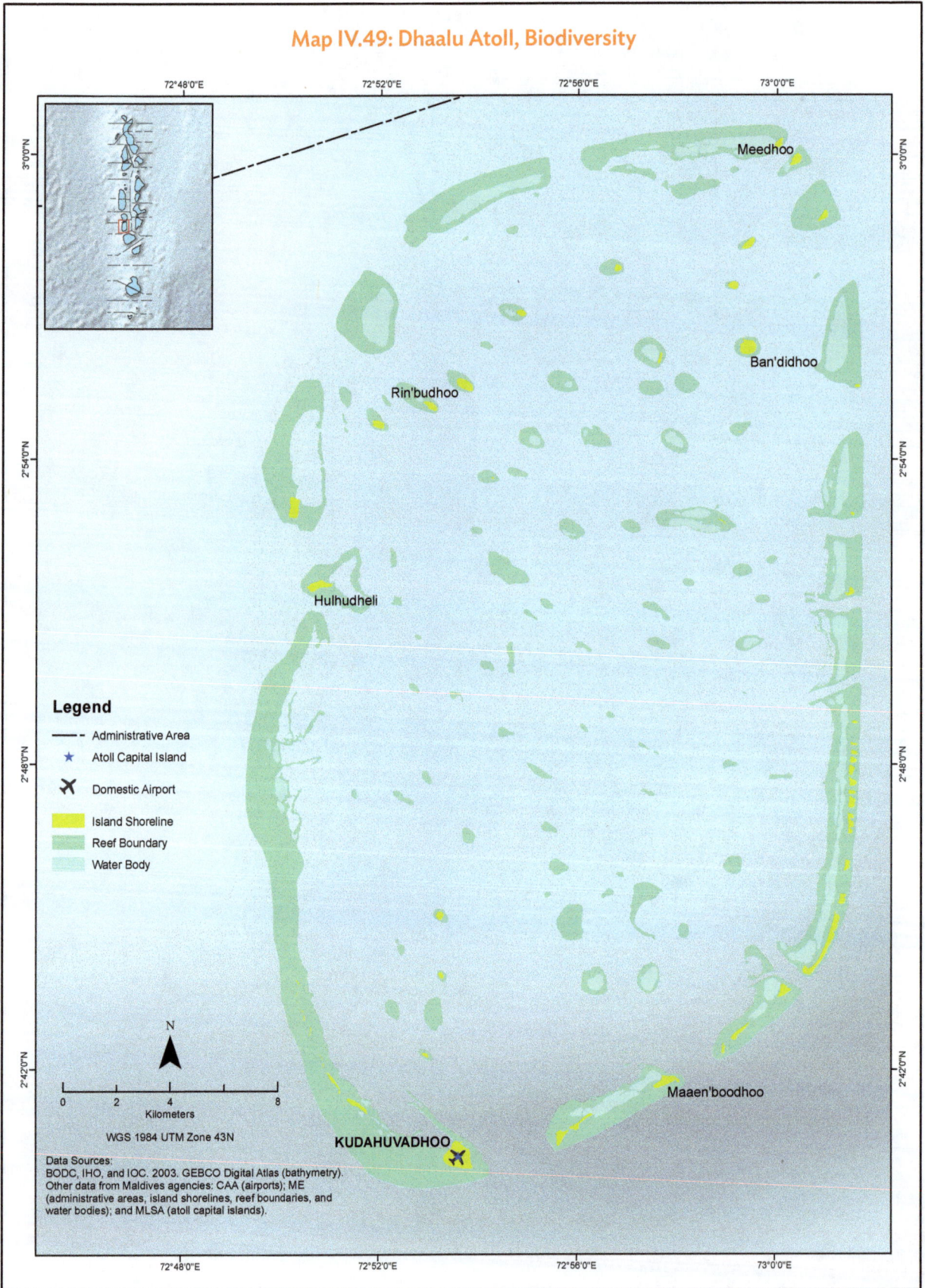

Meedhoo

Ban'didhoo

Rin'budhoo

Hulhudheli

Legend

- – – Administrative Area
- ★ Atoll Capital Island
- ✈ Domestic Airport
- Island Shoreline
- Reef Boundary
- Water Body

N

0   2   4   8
Kilometers

WGS 1984 UTM Zone 43N

Data Sources:
BODC, IHO, and IOC. 2003. GEBCO Digital Atlas (bathymetry).
Other data from Maldives agencies: CAA (airports); ME
(administrative areas, island shorelines, reef boundaries, and
water bodies); and MLSA (atoll capital islands).

Maaen'boodhoo

KUDAHUVADHOO ✈

Map IV.50: Faafu Atoll, Biodiversity

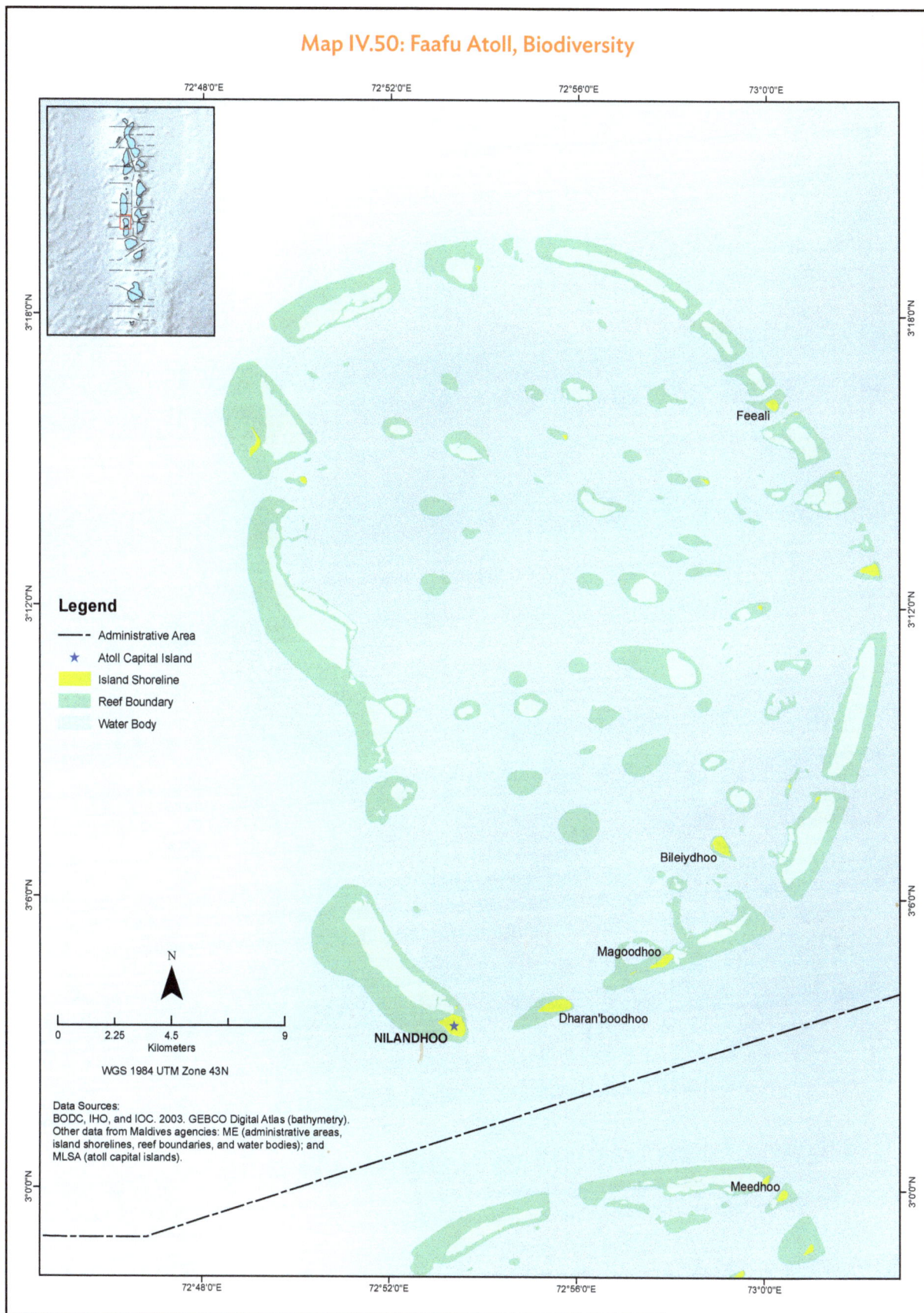

Feeali

Legend

- — - Administrative Area
- ★ Atoll Capital Island
- ▮ Island Shoreline
- ▮ Reef Boundary
- ▮ Water Body

Bileiydhoo

Magoodhoo

Dharan'boodhoo

N

| 0 | 2.25 | 4.5 | | 9 |

Kilometers

WGS 1984 UTM Zone 43N

NILANDHOO

Data Sources:
BODC, IHO, and IOC. 2003. GEBCO Digital Atlas (bathymetry).
Other data from Maldives agencies: ME (administrative areas,
island shorelines, reef boundaries, and water bodies); and
MLSA (atoll capital islands).

Meedhoo

Map IV.51: Gaafu Alifu Atoll, Biodiversity

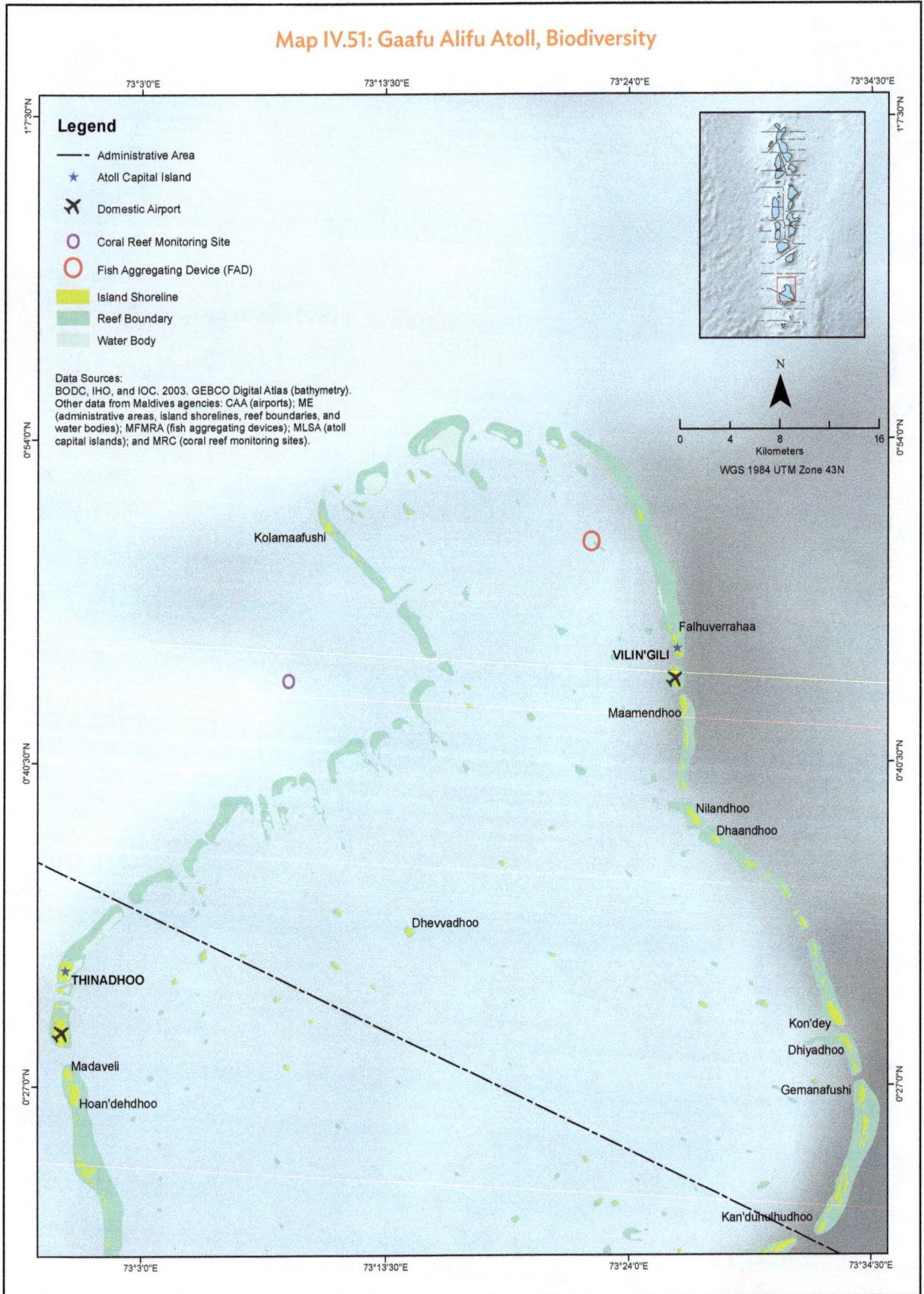

Legend

- – – Administrative Area
- ★ Atoll Capital Island
- ✈ Domestic Airport
- ○ Coral Reef Monitoring Site
- ○ Fish Aggregating Device (FAD)
- Island Shoreline
- Reef Boundary
- Water Body

Data Sources:
BODC, IHO, and IOC. 2003. GEBCO Digital Atlas (bathymetry).
Other data from Maldives agencies: CAA (airports); ME
(administrative areas, island shorelines, reef boundaries, and
water bodies); MFMRA (fish aggregating devices); MLSA (atoll
capital islands); and MRC (coral reef monitoring sites).

N

0   4   8   16
Kilometers
WGS 1984 UTM Zone 43N

Kolamaafushi

Falhuverrahaa
VILIN'GILI
Maamendhoo

Nilandhoo
Dhaandhoo

Dhevvadhoo

THINADHOO

Kon'dey
Dhiyadhoo
Gemanafushi

Madaveli
Hoan'dehdhoo

Kan'duhulhudhoo

Map IV.52: Gaafu Dhaalu Atoll, Biodiversity

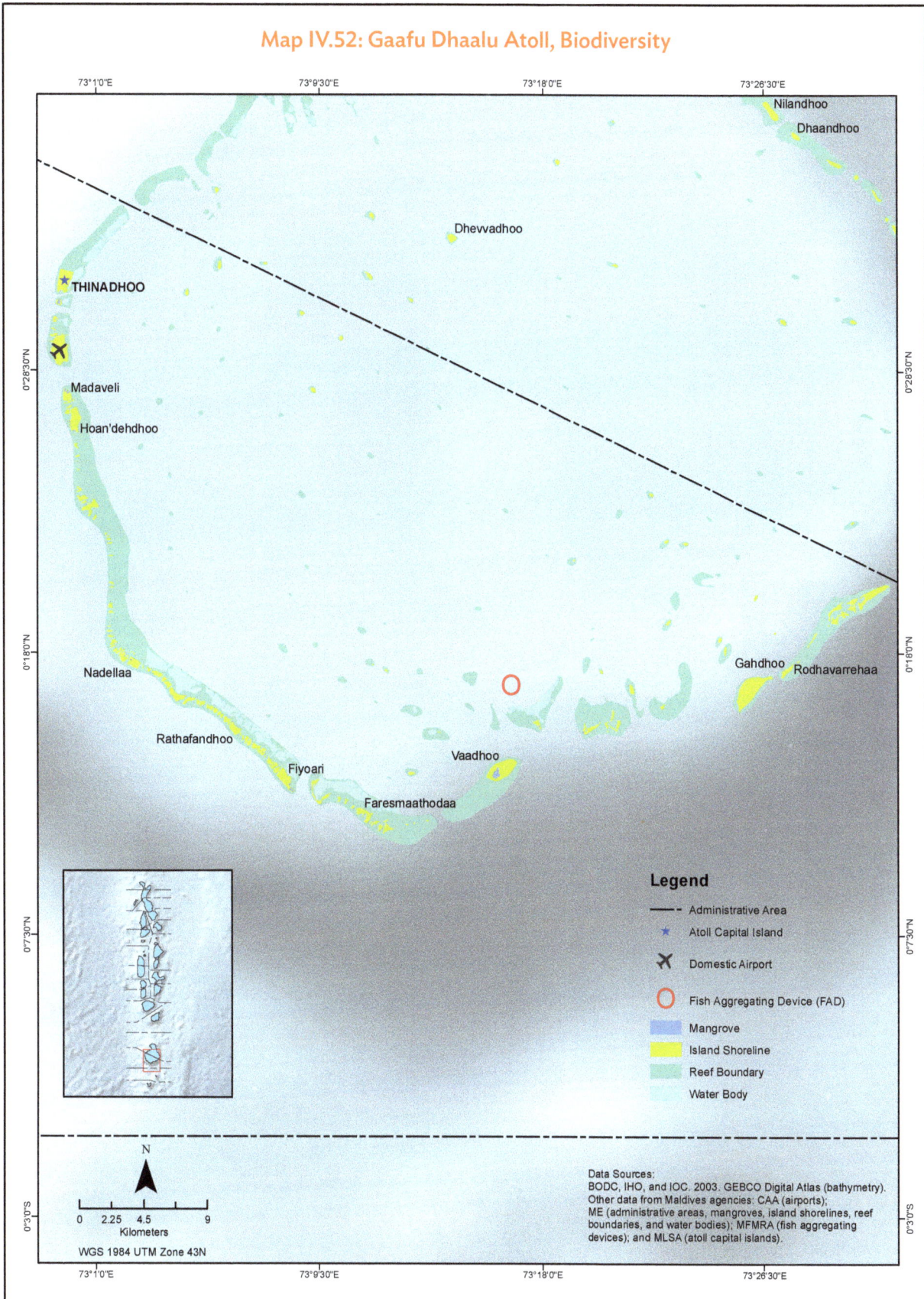

Nilandhoo
Dhaandhoo

Dhevvadhoo

★ THINADHOO

Madaveli

Hoan'dehdhoo

Gahdhoo Rodhavarrehaa

Nadellaa

Rathafandhoo

Fiyoari

Vaadhoo

Faresmaathodaa

Legend

— · — Administrative Area
★ Atoll Capital Island
✈ Domestic Airport
O Fish Aggregating Device (FAD)
▬ Mangrove
▬ Island Shoreline
▬ Reef Boundary
▬ Water Body

N

0   2.25   4.5   9
Kilometers
WGS 1984 UTM Zone 43N

Data Sources:
BODC, IHO, and IOC. 2003. GEBCO Digital Atlas (bathymetry).
Other data from Maldives agencies: CAA (airports);
ME (administrative areas, mangroves, island shorelines, reef
boundaries, and water bodies); MFMRA (fish aggregating
devices); and MLSA (atoll capital islands).

Map IV.53: Gnaviyani Atoll, Biodiversity

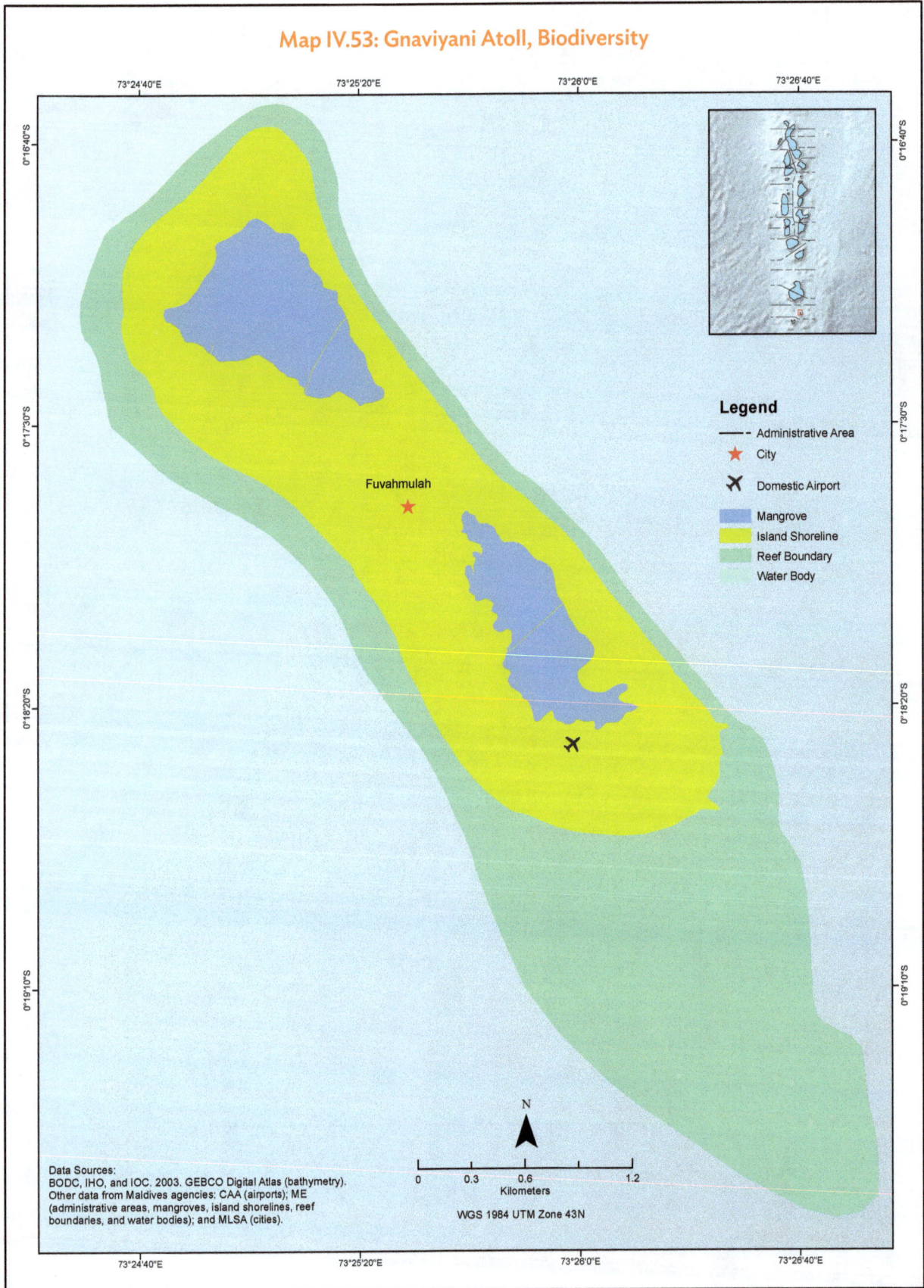

Fuvahmulah

Legend

— — — Administrative Area

★ City

✈ Domestic Airport

▮ Mangrove

▮ Island Shoreline

▮ Reef Boundary

▮ Water Body

Data Sources:
BODC, IHO, and IOC. 2003. GEBCO Digital Atlas (bathymetry).
Other data from Maldives agencies: CAA (airports); ME
(administrative areas, mangroves, island shorelines, reef
boundaries, and water bodies); and MLSA (cities).

N

0   0.3   0.6   1.2
Kilometers

WGS 1984 UTM Zone 43N

Map IV.54: Haa Alifu Atoll, Biodiversity

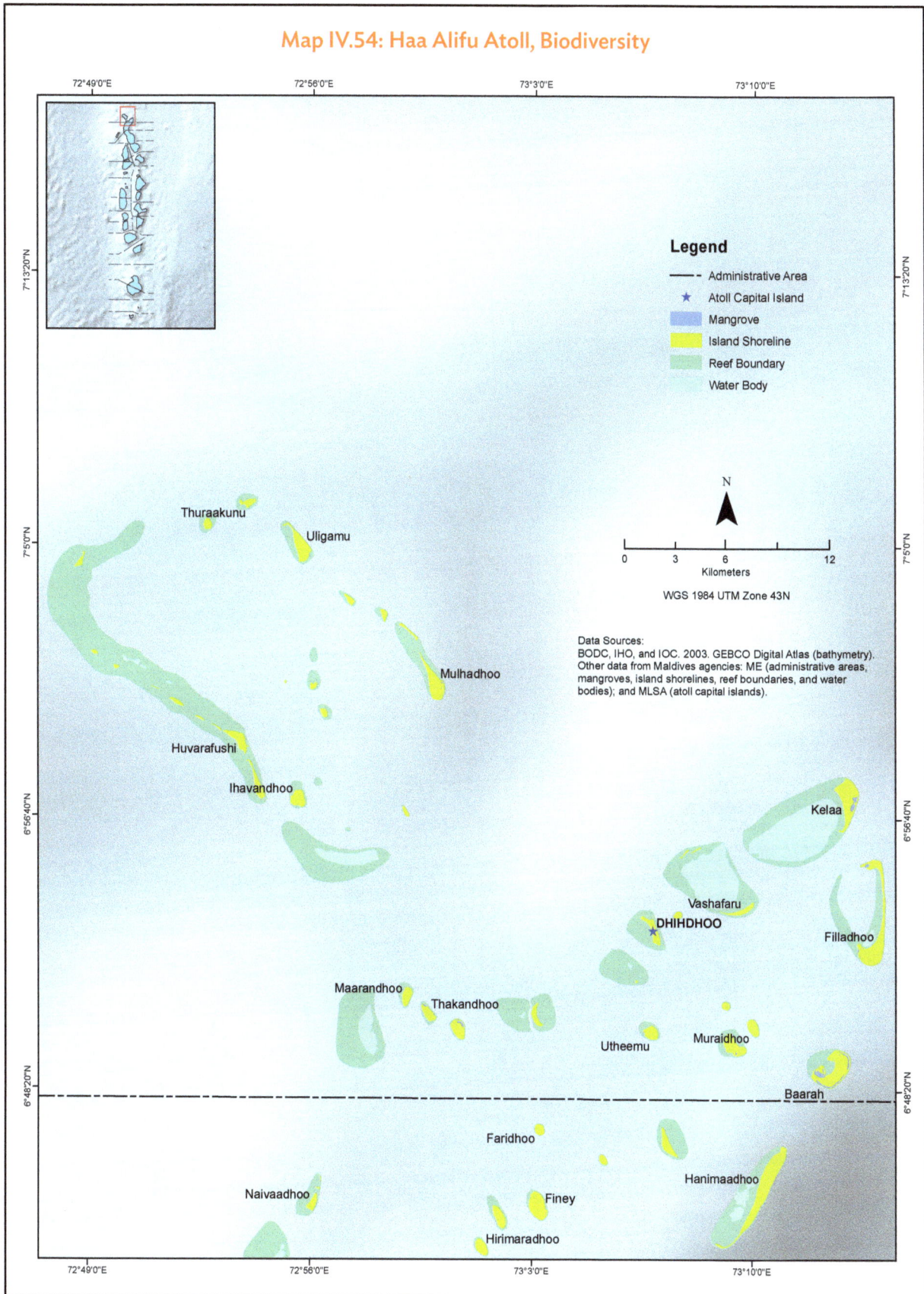

72°49'0"E   72°56'0"E   73°3'0"E   73°10'0"E

Legend
— Administrative Area
★ Atoll Capital Island
Mangrove
Island Shoreline
Reef Boundary
Water Body

N

0   3   6   12
Kilometers
WGS 1984 UTM Zone 43N

Data Sources:
BODC, IHO, and IOC. 2003. GEBCO Digital Atlas (bathymetry).
Other data from Maldives agencies: ME (administrative areas,
mangroves, island shorelines, reef boundaries, and water
bodies); and MLSA (atoll capital islands).

7°13'20"N

7°5'0"N

6°56'40"N

6°48'20"N

Thuraakunu
Uligamu
Mulhadhoo
Huvarafushi
Ihavandhoo
Kelaa
Vashafaru
DHIHDHOO
Filladhoo
Maarandhoo
Thakandhoo
Utheemu
Muraidhoo
Baarah
Faridhoo
Hanimaadhoo
Naivaadhoo
Finey
Hirimaradhoo

Map IV.55: Haa Dhaalu Atoll, Biodiversity

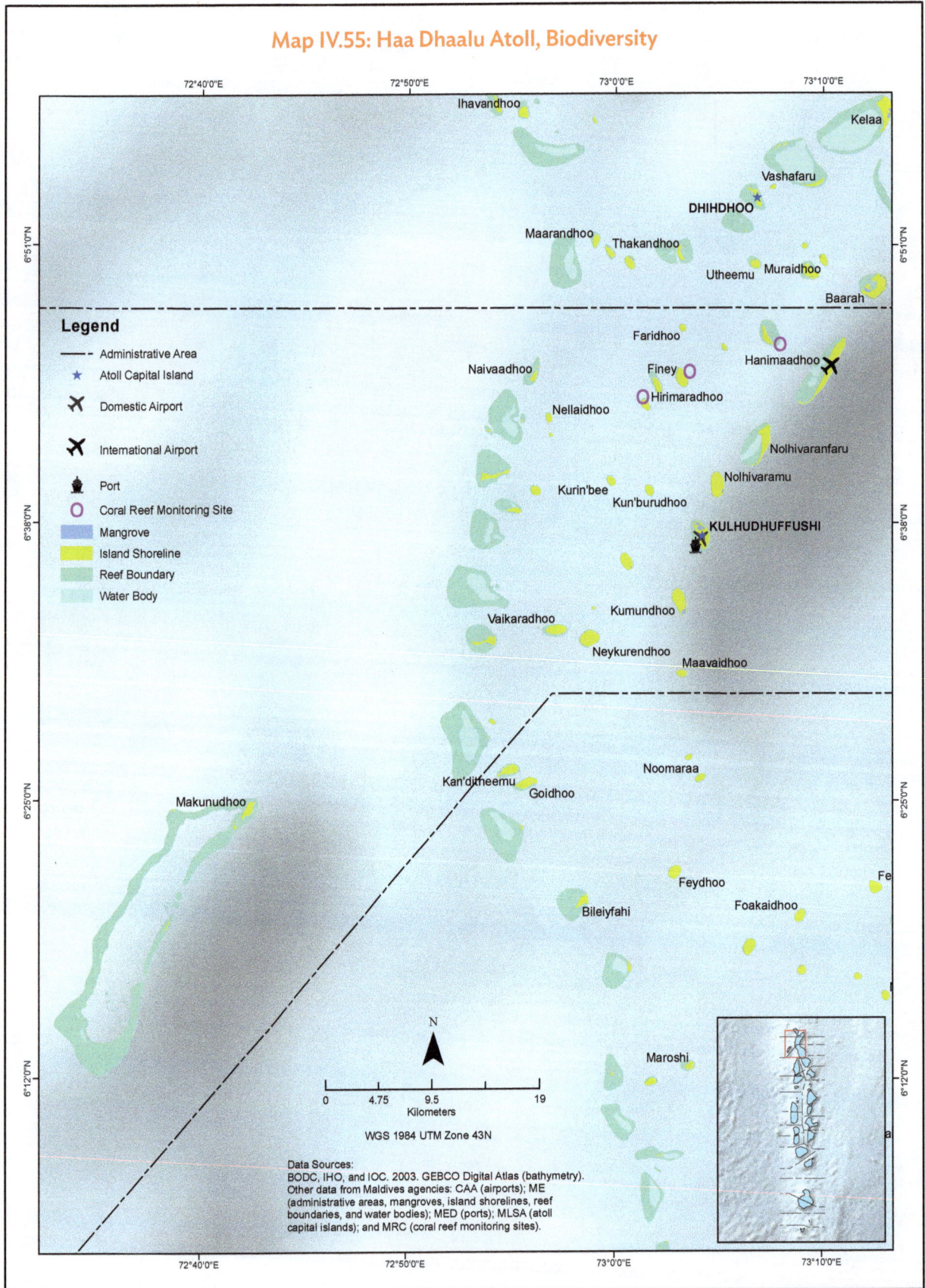

Ihavandhoo

Kelaa

Vashafaru

DHIHDHOO

Maarandhoo   Thakandhoo

Utheemu   Muraidhoo

Baarah

Faridhoo

Naivaadhoo

Finey   Hanimaadhoo

Nellaidhoo   Hirimaradhoo

Nolhivaranfaru

Nolhivaramu

Kurin'bee

Kun'burudhoo

KULHUDHUFFUSHI

Kumundhoo

Vaikaradhoo

Neykurendhoo

Maavaidhoo

Legend

- —— Administrative Area
- ★ Atoll Capital Island
- ✕ Domestic Airport
- ✈ International Airport
- ⚓ Port
- ○ Coral Reef Monitoring Site
- ▮ Mangrove
- ▮ Island Shoreline
- ▮ Reef Boundary
- ▮ Water Body

Noomaraa

Kan'ditheemu   Goidhoo

Makunudhoo

Feydhoo

Foakaidhoo

Fe

Bileiyfahi

Maroshi

N

0   4.75   9.5   19
Kilometers

WGS 1984 UTM Zone 43N

Data Sources:
BODC, IHO, and IOC. 2003. GEBCO Digital Atlas (bathymetry).
Other data from Maldives agencies: CAA (airports); ME
(administrative areas, mangroves, island shorelines, reef
boundaries, and water bodies); MED (ports); MLSA (atoll
capital islands); and MRC (coral reef monitoring sites).

Map IV.56: Laamu Atoll, Biodiversity

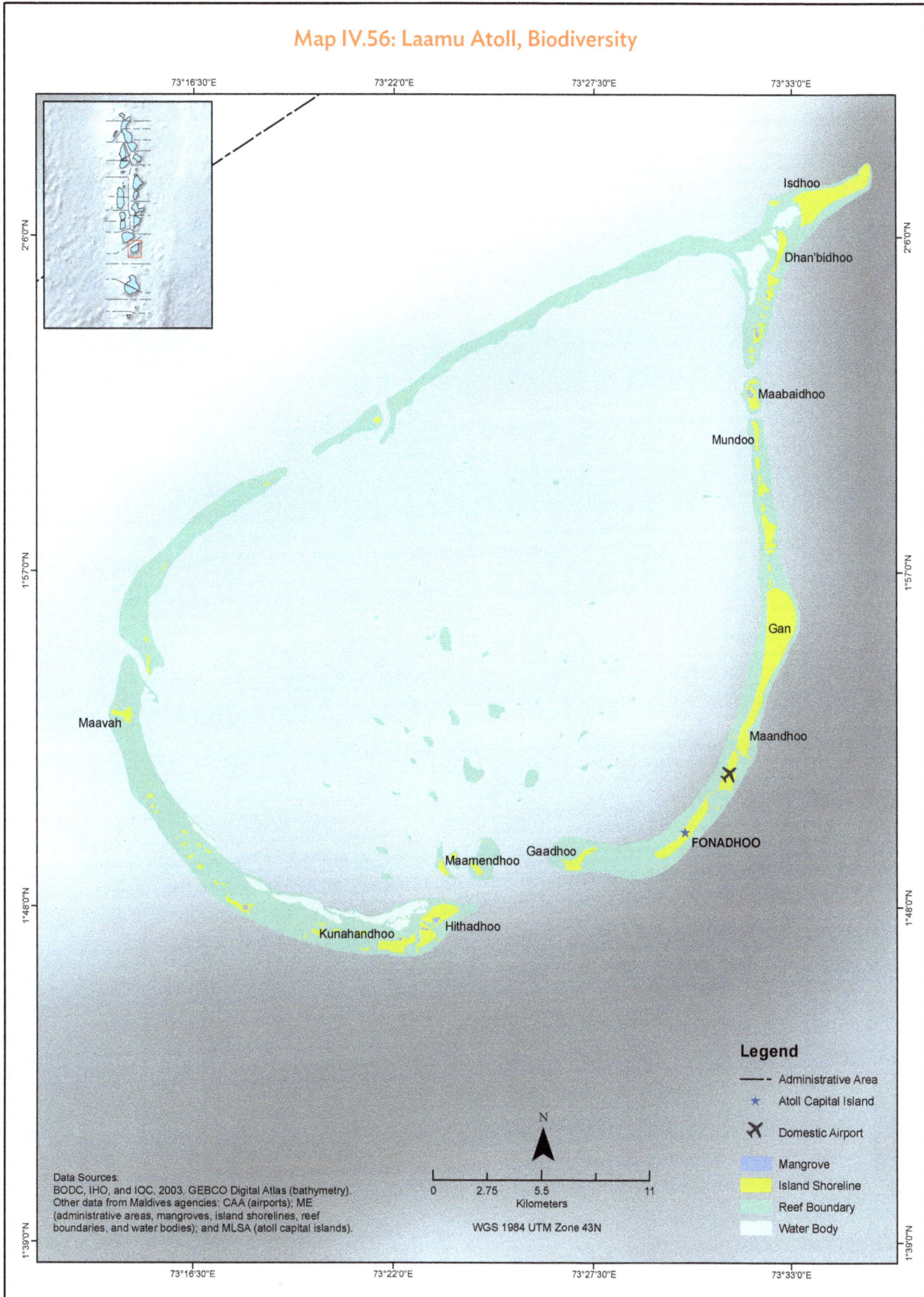

Isdhoo

Dhan'bidhoo

Maabaidhoo

Mundoo

Gan

Maandhoo

FONADHOO

Maavah

Gaadhoo

Maamendhoo

Kunahandhoo Hithadhoo

N

0   2.75   5.5   11
Kilometers

WGS 1984 UTM Zone 43N

Data Sources:
BODC, IHO, and IOC. 2003. GEBCO Digital Atlas (bathymetry).
Other data from Maldives agencies: CAA (airports); ME
(administrative areas, mangroves, island shorelines, reef
boundaries, and water bodies); and MLSA (atoll capital islands).

Legend
- - - Administrative Area
★ Atoll Capital Island
✈ Domestic Airport
Mangrove
Island Shoreline
Reef Boundary
Water Body

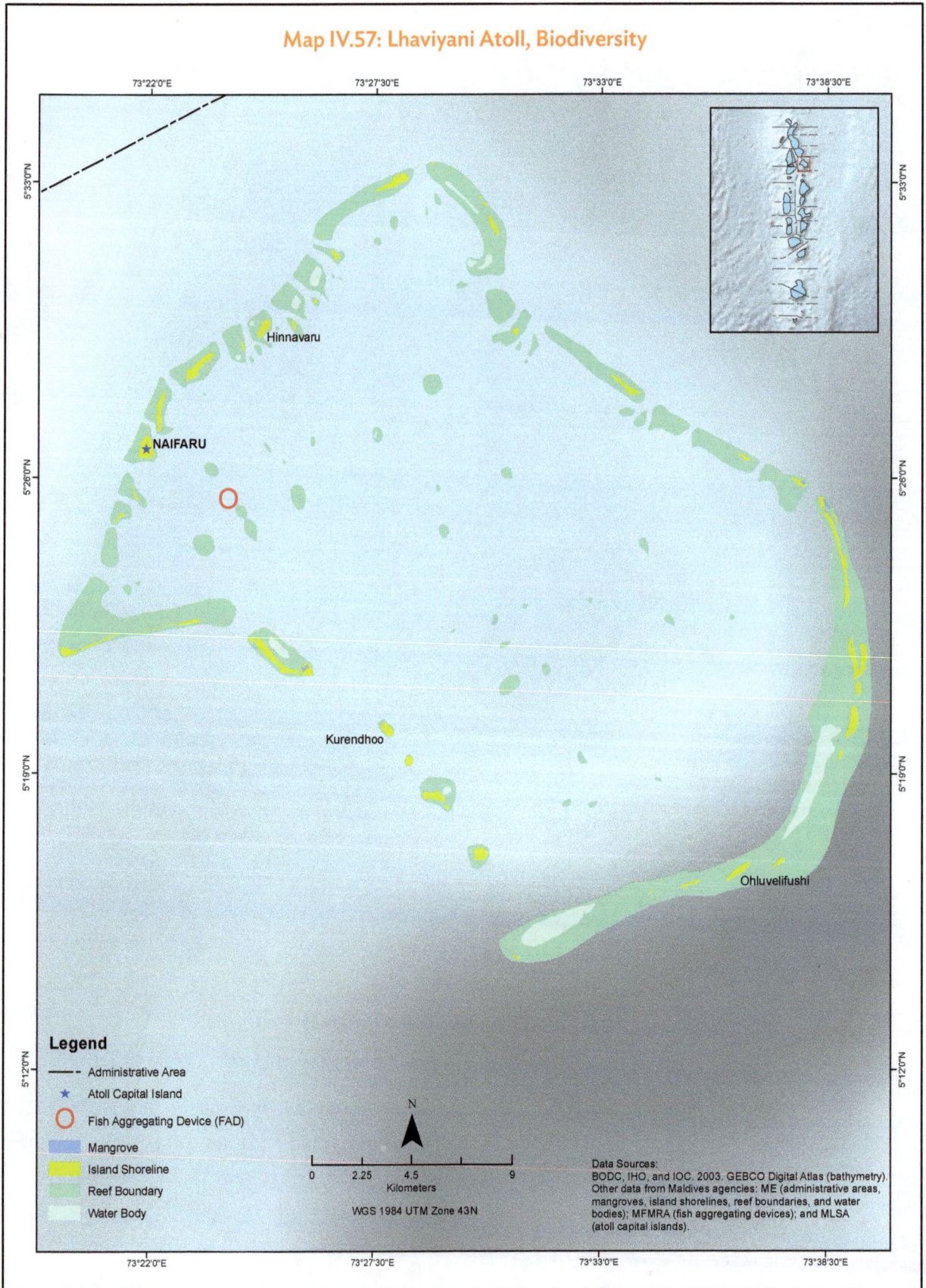

Map IV.57: Lhaviyani Atoll, Biodiversity

Hinnavaru

★NAIFARU

Kurendhoo

Ohluvelifushi

Legend

— — Administrative Area

★ Atoll Capital Island

◯ Fish Aggregating Device (FAD)

Mangrove

Island Shoreline

Reef Boundary

Water Body

N

0   2.25   4.5   9
Kilometers

WGS 1984 UTM Zone 43N

Data Sources:
BODC, IHO, and IOC. 2003. GEBCO Digital Atlas (bathymetry).
Other data from Maldives agencies: ME (administrative areas,
mangroves, island shorelines, reef boundaries, and water
bodies); MFMRA (fish aggregating devices); and MLSA
(atoll capital islands).

Map IV.58: Meemu Atoll, Biodiversity

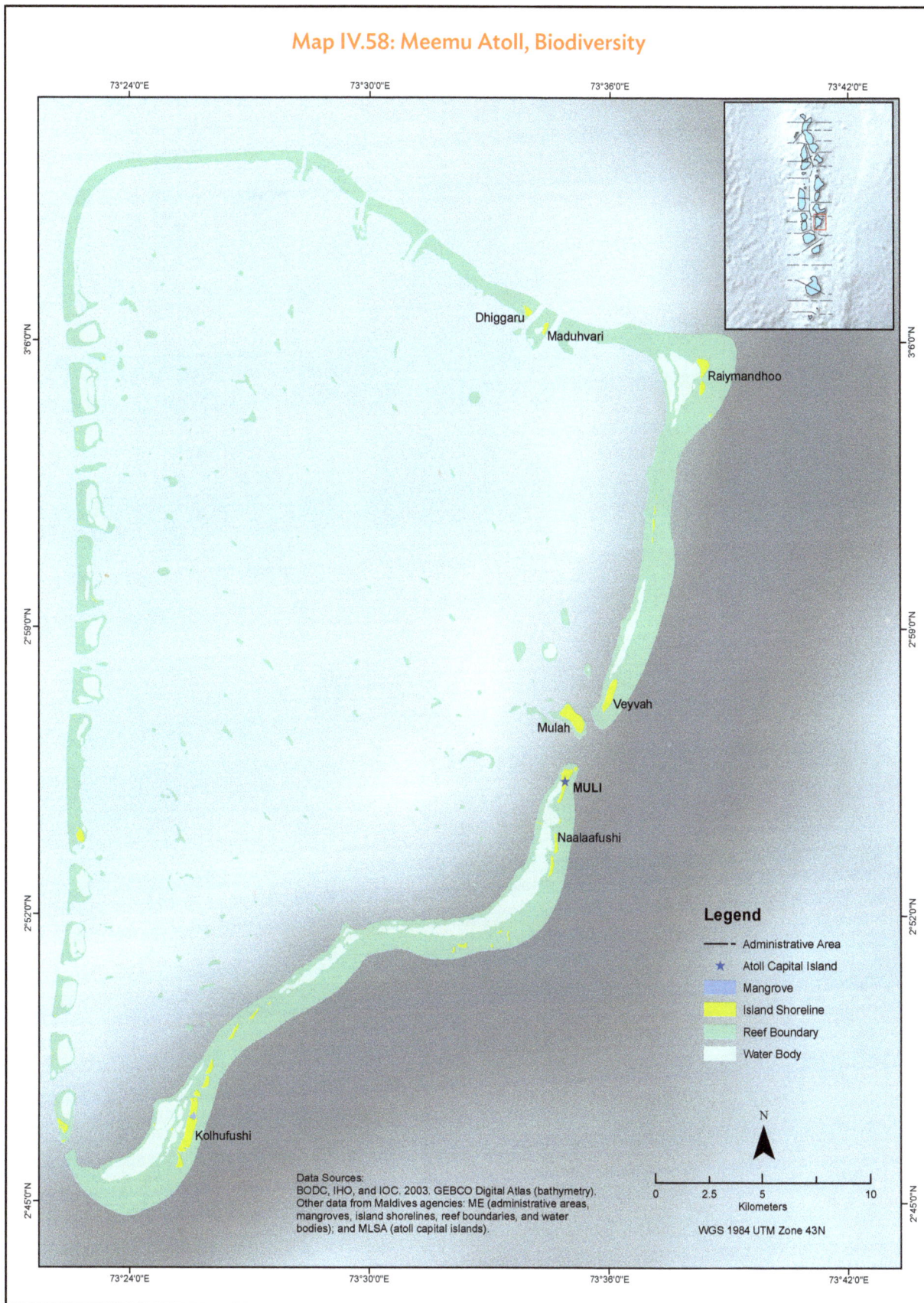

73°24'0"E   73°30'0"E   73°36'0"E   73°42'0"E

Dhiggaru
Maduhvari
Raiymandhoo

Veyvah
Mulah

★ MULI

Naalaafushi

Kolhufushi

Legend

— Administrative Area
★ Atoll Capital Island
   Mangrove
   Island Shoreline
   Reef Boundary
   Water Body

N

Data Sources:
BODC, IHO, and IOC. 2003. GEBCO Digital Atlas (bathymetry).
Other data from Maldives agencies: ME (administrative areas,
mangroves, island shorelines, reef boundaries, and water
bodies); and MLSA (atoll capital islands).

0   2.5   5   10
Kilometers

WGS 1984 UTM Zone 43N

Map IV.59: Noonu Atoll, Biodiversity

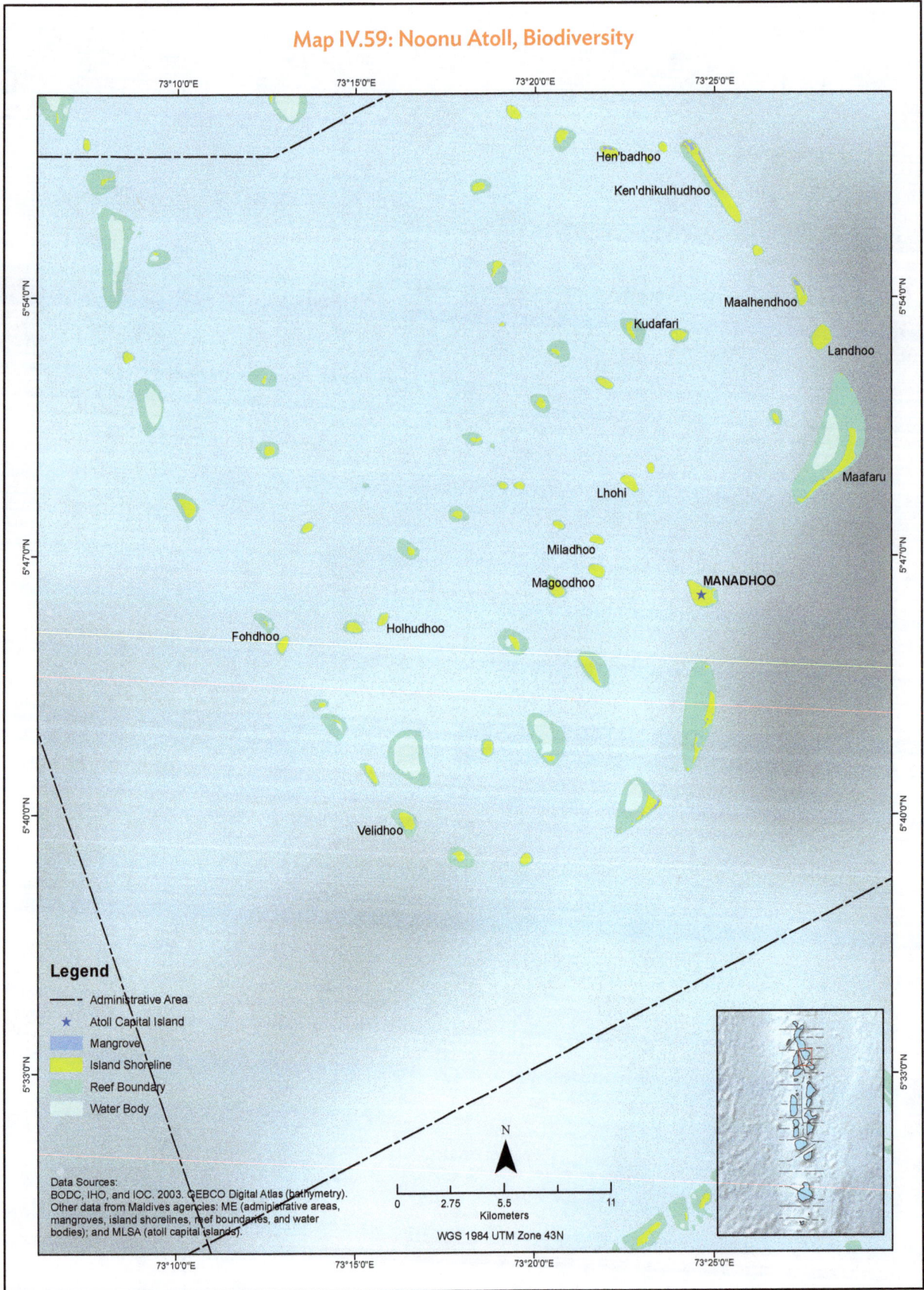

73°10'0"E   73°15'0"E   73°20'0"E   73°25'0"E

Hen'badhoo

Ken'dhikulhudhoo

5°54'0"N

Maalhendhoo

Kudafari

Landhoo

Maafaru

Lhohi

Miladhoo

5°47'0"N

Magoodhoo   MANADHOO

Holhudhoo

Fohdhoo

Velidhoo

5°40'0"N

Legend

— Administrative Area
★ Atoll Capital Island
▮ Mangrove
▮ Island Shoreline
▮ Reef Boundary
▮ Water Body

Data Sources:
BODC, IHO, and IOC. 2003. GEBCO Digital Atlas (bathymetry).
Other data from Maldives agencies: ME (administrative areas,
mangroves, island shorelines, reef boundaries, and water
bodies); and MLSA (atoll capital islands).

N

0   2.75   5.5   11
Kilometers

WGS 1984 UTM Zone 43N

Map IV.60: North Malé Atoll, Biodiversity

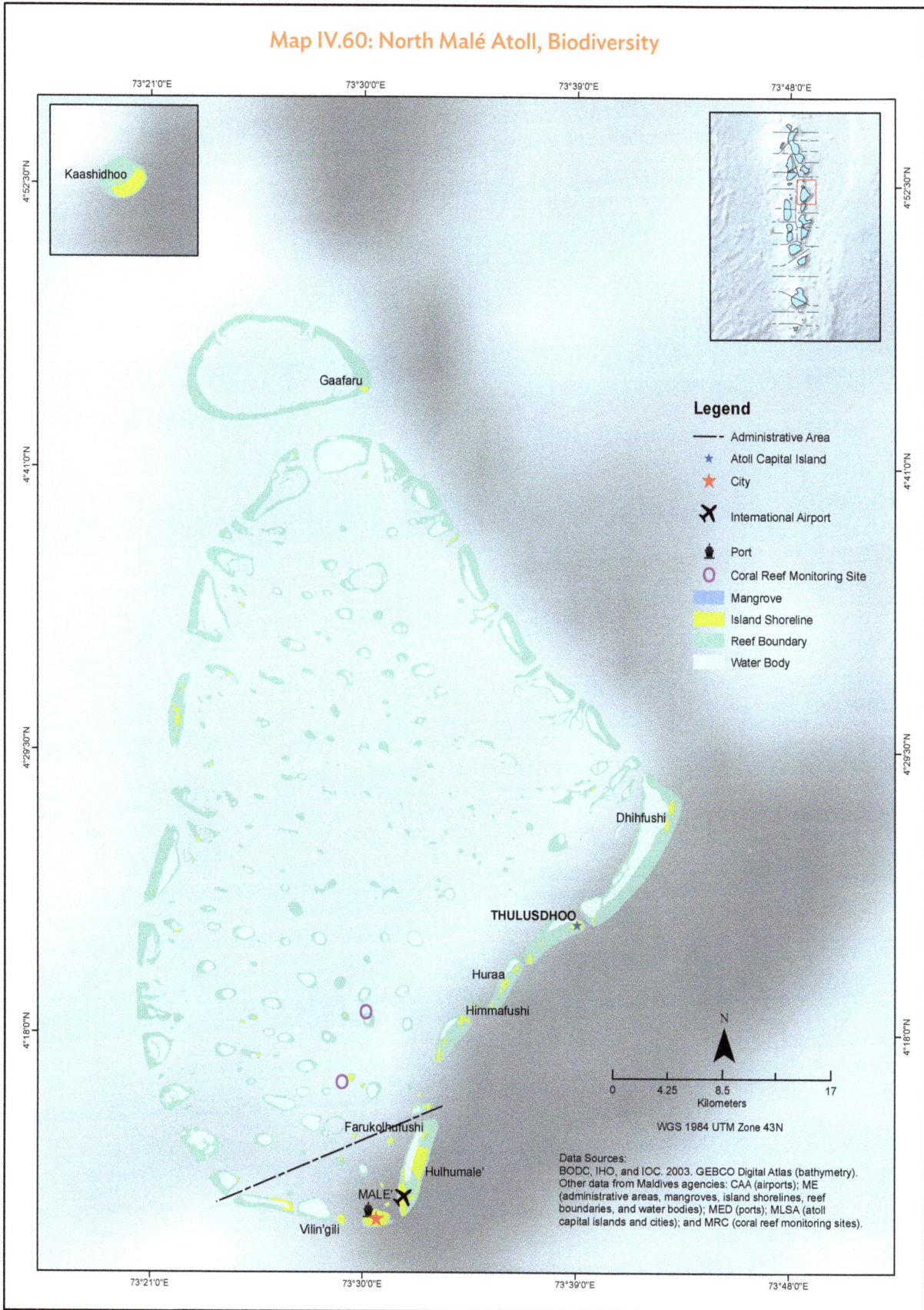

Kaashidhoo

Gaafaru

Dhihfushi

THULUSDHOO

Huraa

Himmafushi

Farukolhufushi

Hulhumale'

MALE'

Vilin'gili

Legend

— — Administrative Area
★ Atoll Capital Island
★ City
✈ International Airport
⚓ Port
○ Coral Reef Monitoring Site
▮ Mangrove
▮ Island Shoreline
▮ Reef Boundary
▯ Water Body

N

0   4.25   8.5   17
Kilometers

WGS 1984 UTM Zone 43N

Data Sources:
BODC, IHO, and IOC. 2003. GEBCO Digital Atlas (bathymetry).
Other data from Maldives agencies: CAA (airports); ME
(administrative areas, mangroves, island shorelines, reef
boundaries, and water bodies); MED (ports); MLSA (atoll
capital islands and cities); and MRC (coral reef monitoring sites).

Map IV.61: Raa Atoll, Biodiversity

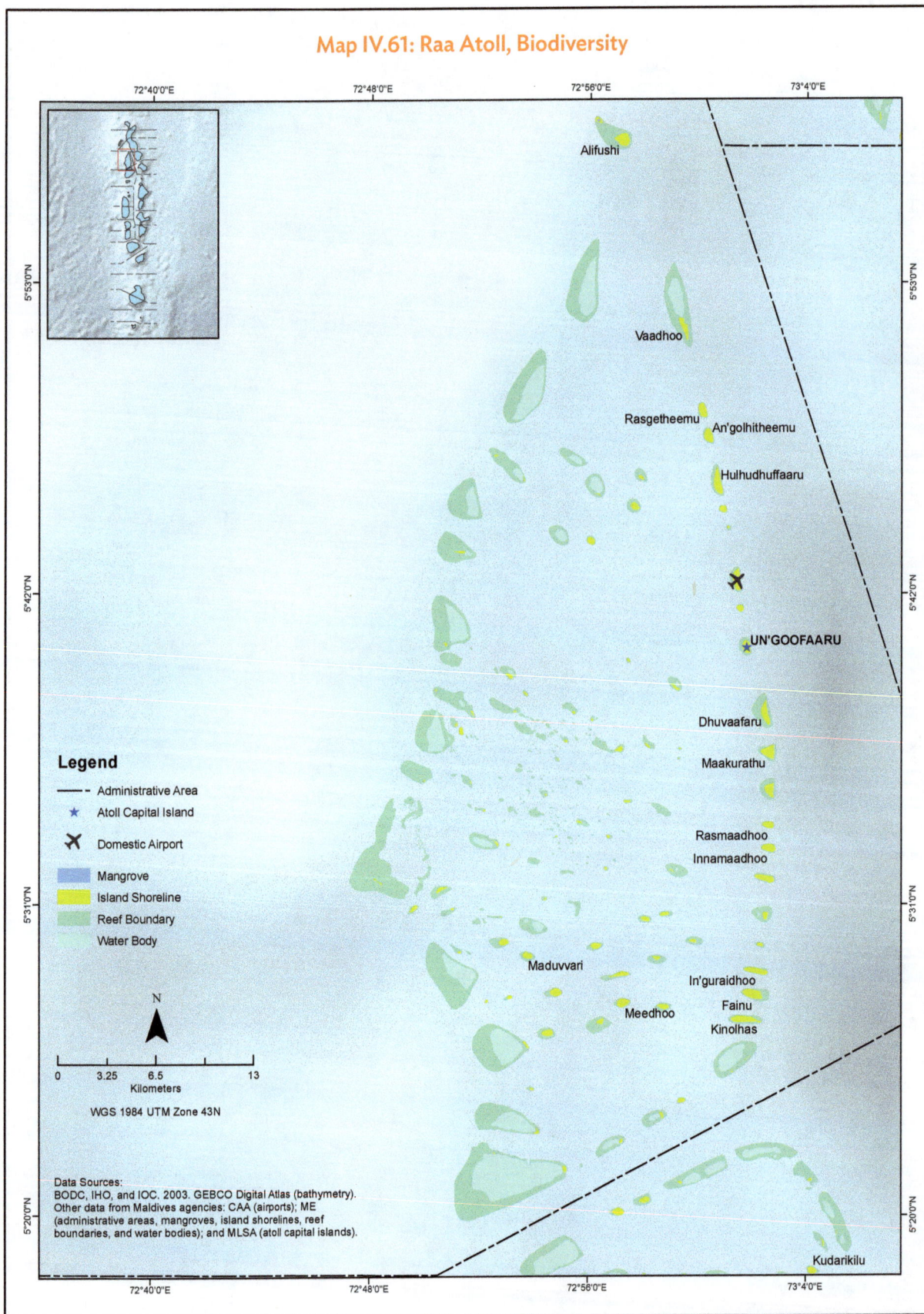

Alifushi

Vaadhoo

Rasgetheemu   An'golhitheemu

Hulhudhuffaaru

UN'GOOFAARU

Dhuvaafaru

Maakurathu

Rasmaadhoo
Innamaadhoo

Maduvvari

In'guraidhoo

Meedhoo   Fainu

Kinolhas

Kudarikilu

Legend

— — — Administrative Area

★ Atoll Capital Island

✈ Domestic Airport

Mangrove

Island Shoreline

Reef Boundary

Water Body

N

| 0 | 3.25 | 6.5 | | 13 |
Kilometers

WGS 1984 UTM Zone 43N

Data Sources:
BODC, IHO, and IOC. 2003. GEBCO Digital Atlas (bathymetry).
Other data from Maldives agencies: CAA (airports); ME
(administrative areas, mangroves, island shorelines, reef
boundaries, and water bodies); and MLSA (atoll capital islands).

Map IV.62: Shaviyani Atoll, Biodiversity

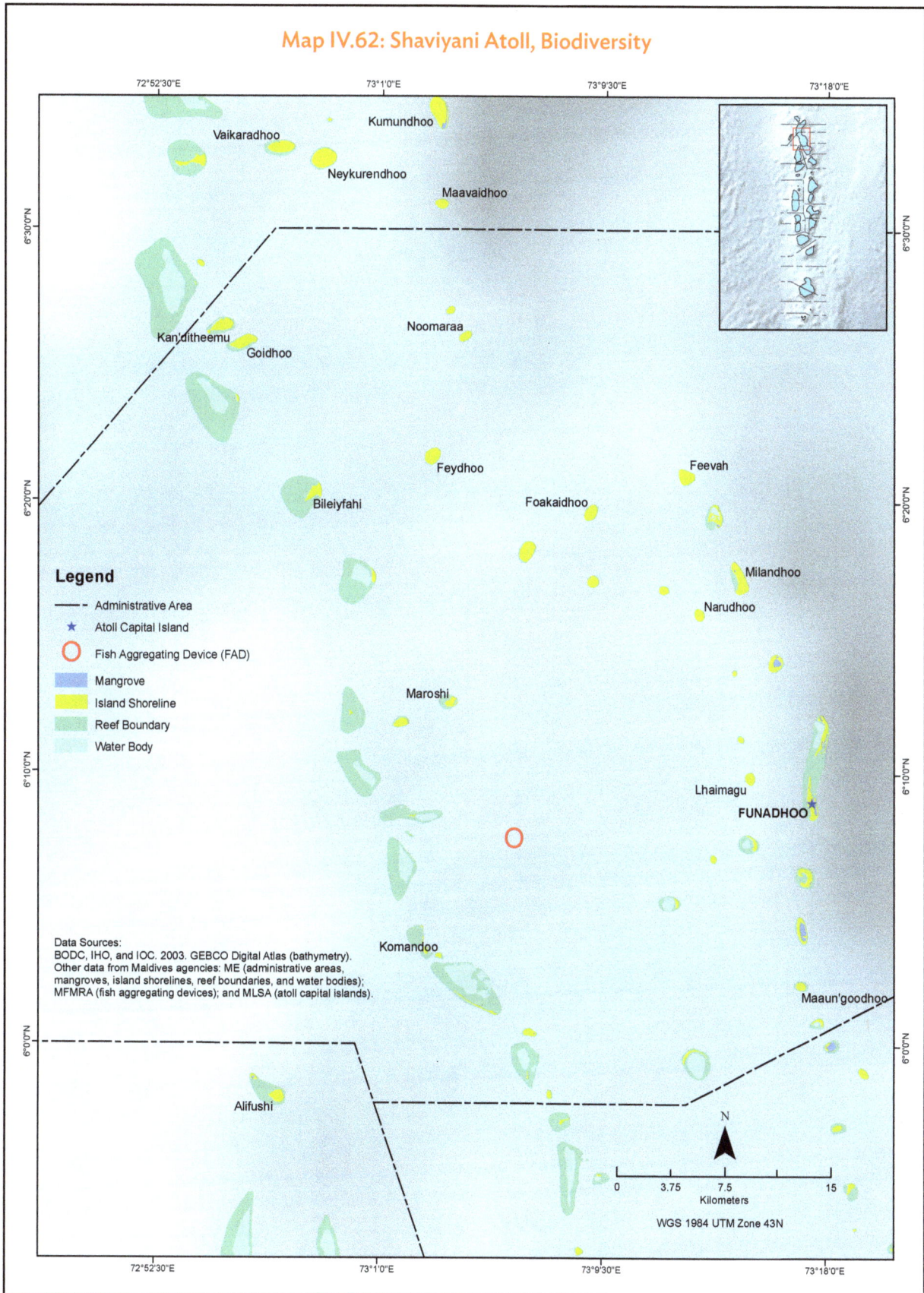

Kumundhoo
Vaikaradhoo
Neykurendhoo
Maavaidhoo
Noomaraa
Kan'ditheemu
Goidhoo
Feydhoo
Feevah
Bileiyfahi
Foakaidhoo
Milandhoo
Narudhoo
Maroshi
Lhaimagu
FUNADHOO
Komandoo
Maaun'goodhoo
Alifushi

Legend
- — Administrative Area
- ★ Atoll Capital Island
- ◯ Fish Aggregating Device (FAD)
- Mangrove
- Island Shoreline
- Reef Boundary
- Water Body

Data Sources:
BODC, IHO, and IOC. 2003. GEBCO Digital Atlas (bathymetry).
Other data from Maldives agencies: ME (administrative areas,
mangroves, island shorelines, reef boundaries, and water bodies);
MFMRA (fish aggregating devices); and MLSA (atoll capital islands).

N

0   3.75   7.5   15
Kilometers

WGS 1984 UTM Zone 43N

Map IV.63: South Malé Atoll, Biodiversity

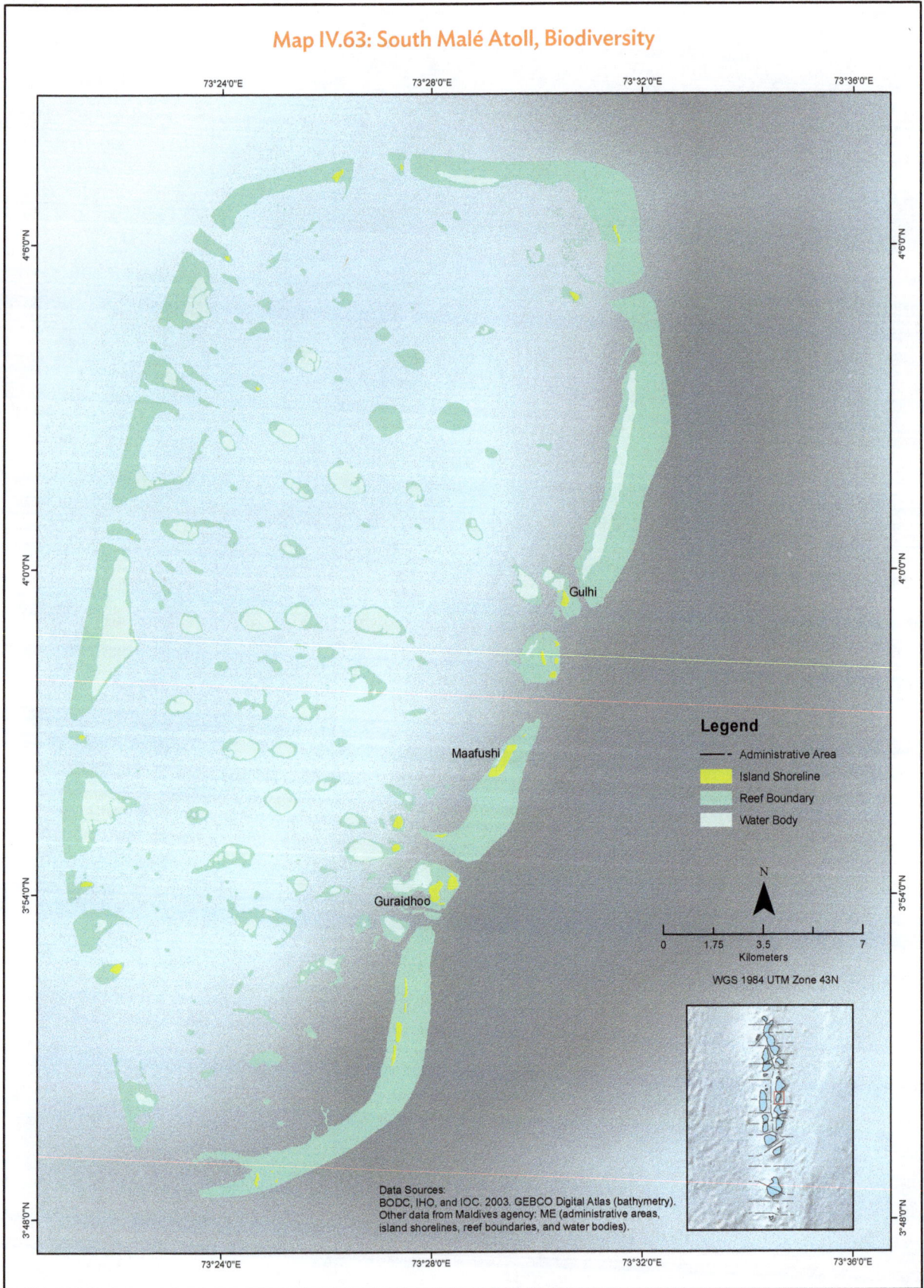

Gulhi

Maafushi

Guraidhoo

Legend

- – – Administrative Area
- Island Shoreline
- Reef Boundary
- Water Body

N

0   1.75   3.5   7
Kilometers

WGS 1984 UTM Zone 43N

Data Sources:
BODC, IHO, and IOC. 2003. GEBCO Digital Atlas (bathymetry).
Other data from Maldives agency: ME (administrative areas,
island shorelines, reef boundaries, and water bodies).

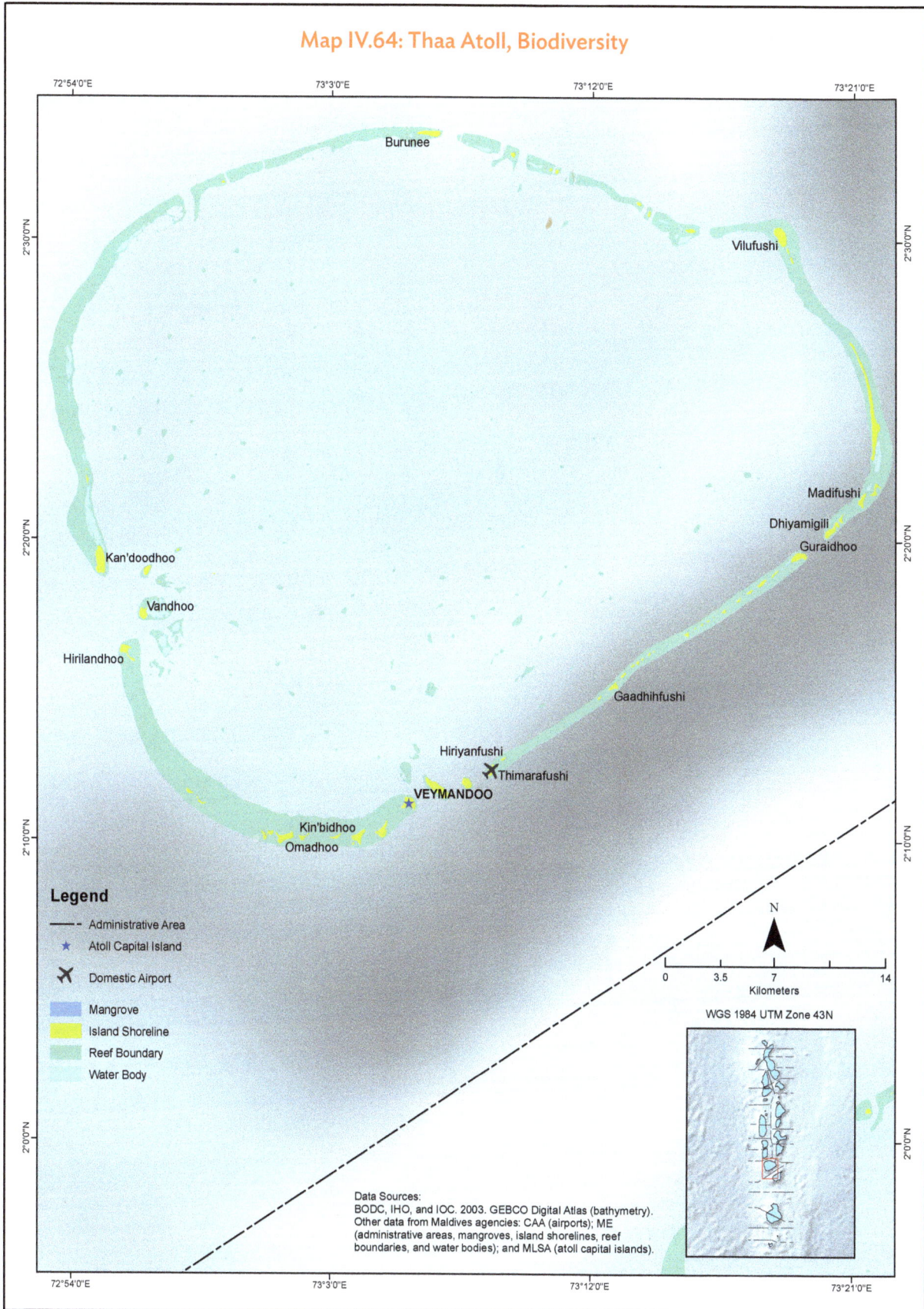

Map IV.64: Thaa Atoll, Biodiversity

Burunee

Vilufushi

Madifushi

Dhiyamigili

Guraidhoo

Kan'doodhoo

Vandhoo

Hirilandhoo

Gaadhihfushi

Hiriyanfushi
Thimarafushi
VEYMANDOO

Kin'bidhoo
Omadhoo

Legend

- Administrative Area
★ Atoll Capital Island
✕ Domestic Airport
Mangrove
Island Shoreline
Reef Boundary
Water Body

N

0 3.5 7 14
Kilometers

WGS 1984 UTM Zone 43N

Data Sources:
BODC, IHO, and IOC. 2003. GEBCO Digital Atlas (bathymetry).
Other data from Maldives agencies: CAA (airports); ME
(administrative areas, mangroves, island shorelines, reef
boundaries, and water bodies); and MLSA (atoll capital islands).

Map IV.65: Vaavu Atoll, Biodiversity

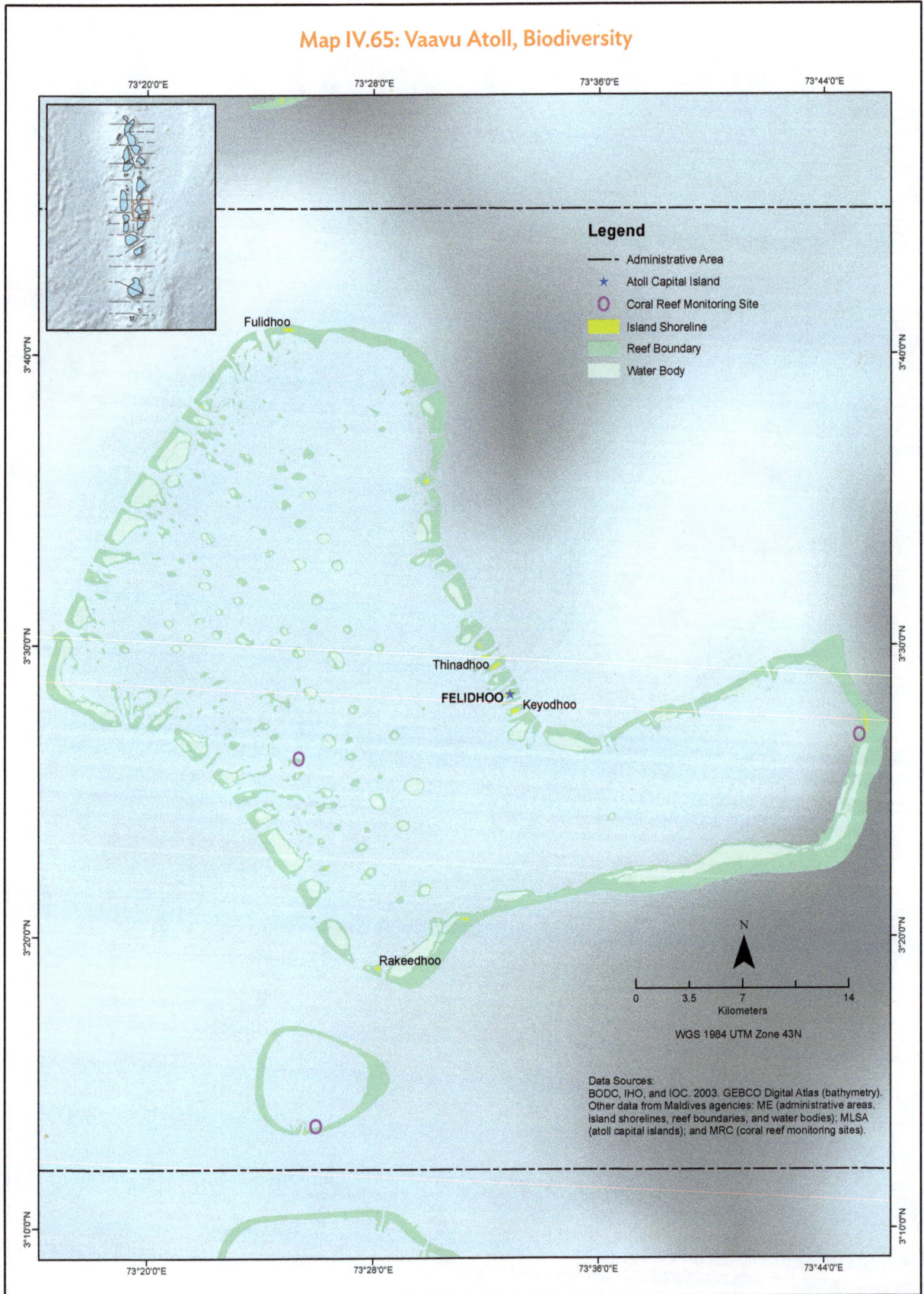

73°20'0"E   73°28'0"E   73°36'0"E   73°44'0"E

Legend

‑‑‑‑ Administrative Area

★ Atoll Capital Island

⬡ Coral Reef Monitoring Site

▮ Island Shoreline

▮ Reef Boundary

▮ Water Body

Fulidhoo

3°40'0"N   3°40'0"N

Thinadhoo

3°30'0"N   3°30'0"N

FELIDHOO ★ Keyodhoo

3°20'0"N   3°20'0"N

Rakeedhoo

N

0   3.5   7   14

Kilometers

WGS 1984 UTM Zone 43N

Data Sources:
BODC, IHO, and IOC. 2003. GEBCO Digital Atlas (bathymetry).
Other data from Maldives agencies: ME (administrative areas,
island shorelines, reef boundaries, and water bodies); MLSA
(atoll capital islands); and MRC (coral reef monitoring sites).

3°10'0"N   3°10'0"N

73°20'0"E   73°28'0"E   73°36'0"E   73°44'0"E

Threats to Marine and Coastal Biodiversity

Coastal Erosion

The islands of Maldives, with an average elevation of 1.4 meters above sea level, are prone to coastal erosions and inundation. Natural factors such as tides, waves, and surges cause these coastal erosions. However, human activities such as sand mining increase the severity of beach erosion. The rising global mean sea level is another threatening factor. Global mean sea level, which is connected to rising temperature, would increase the country's coastal erosion. As of 2017, 45 islands are very severely eroded, 20 are severely eroded, and 18 are slightly eroded. Small island resorts are greatly vulnerable to coastal erosions and, as a result, are already losing economic gains (Emerton, Baig, and Saleem 2009).

Coastal erosion. Wave breakers are installed along the coast to protect the beach from erosion (photo by Erwin Österreich).

Map IV.66: Addu City, Coastal Erosion

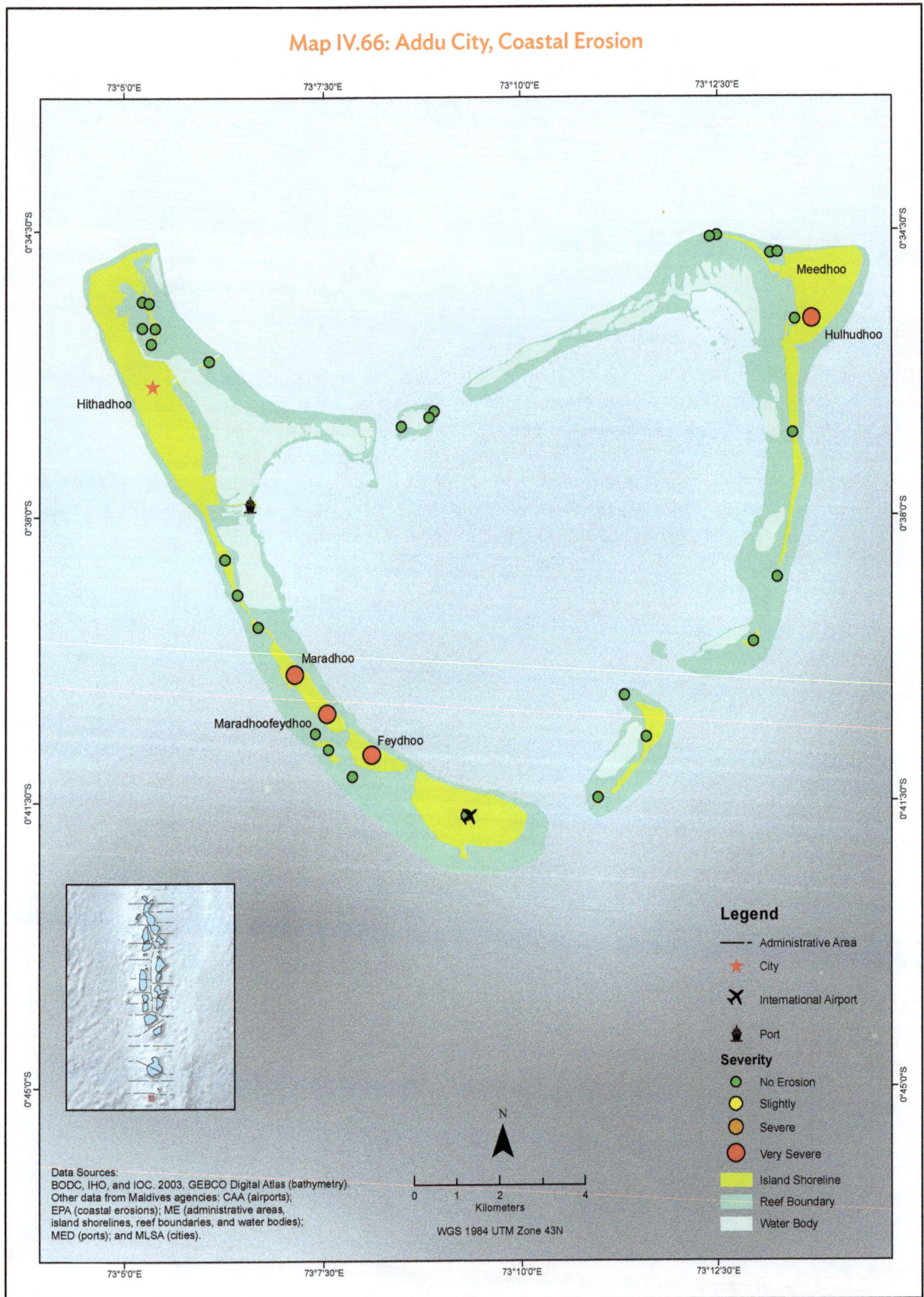

Meedhoo

Hulhudhoo

Hithadhoo

Maradhoo

Maradhoofeydhoo

Feydhoo

Legend

— — — Administrative Area

★ City

✈ International Airport

⛩ Port

Severity

● No Erosion

● Slightly

● Severe

● Very Severe

▮ Island Shoreline

▮ Reef Boundary

▮ Water Body

Data Sources:
BODC, IHO, and IOC. 2003. GEBCO Digital Atlas (bathymetry).
Other data from Maldives agencies: CAA (airports);
EPA (coastal erosions); ME (administrative areas,
island shorelines, reef boundaries, and water bodies);
MED (ports); and MLSA (cities).

N

0   1   2   3   4
Kilometers

WGS 1984 UTM Zone 43N

Map IV.67: Alifu Alifu Atoll, Coastal Erosion

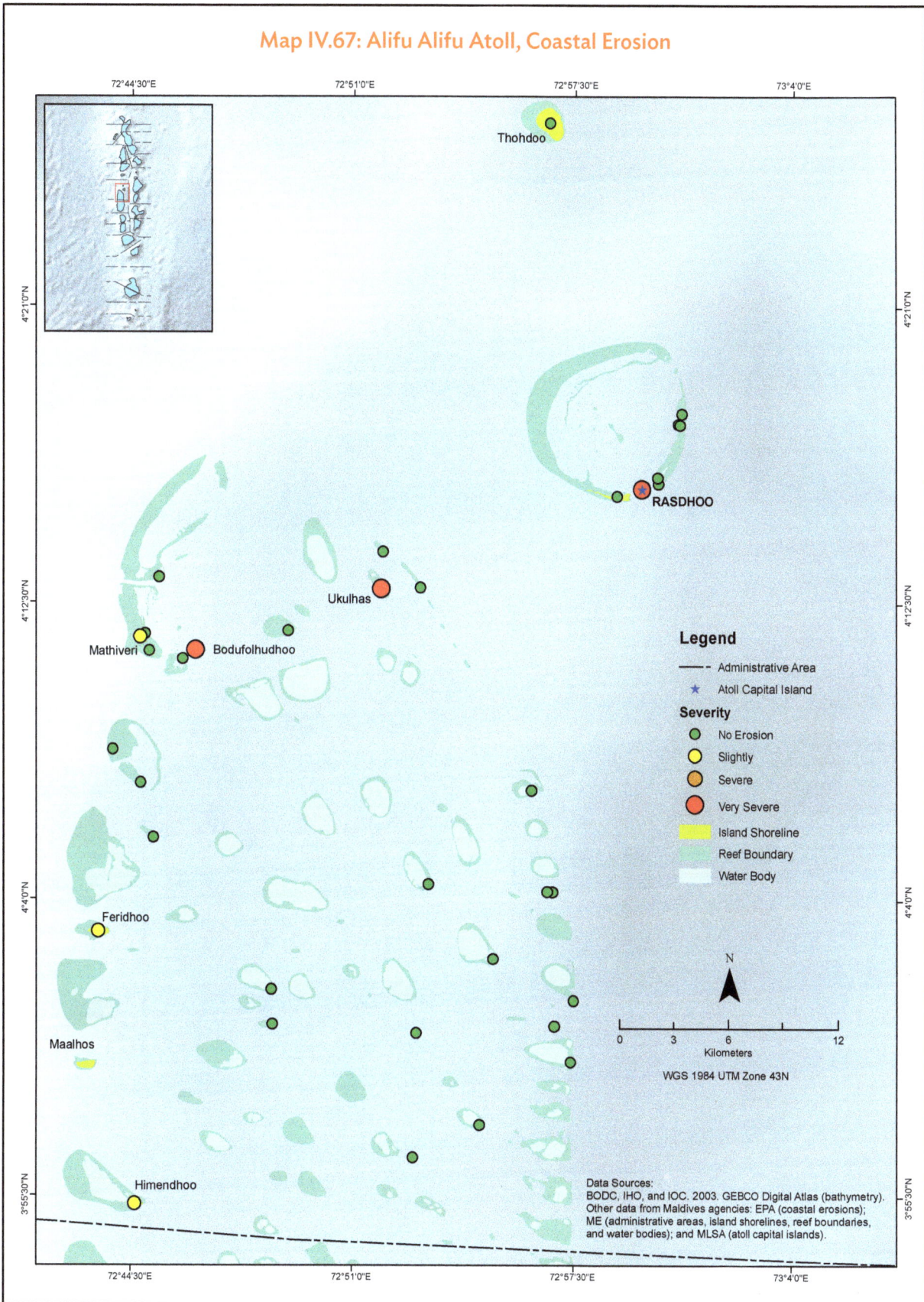

72°44'30"E   72°51'0"E   72°57'30"E   73°4'0"E

4°21'0"N

Thohdoo

RASDHOO

4°12'30"N

Ukulhas

Mathiveri   Bodufolhudhoo

Legend

— — — Administrative Area

★ Atoll Capital Island

Severity

● No Erosion
● Slightly
● Severe
● Very Severe

▮ Island Shoreline
▮ Reef Boundary
▮ Water Body

4°4'0"N

Feridhoo

N

Maalhos

0   3   6   12
Kilometers

WGS 1984 UTM Zone 43N

3°55'30"N

Himendhoo

Data Sources:
BODC, IHO, and IOC. 2003. GEBCO Digital Atlas (bathymetry).
Other data from Maldives agencies: EPA (coastal erosions);
ME (administrative areas, island shorelines, reef boundaries,
and water bodies); and MLSA (atoll capital islands).

72°44'30"E   72°51'0"E   72°57'30"E   73°4'0"E

Map IV.68: Alifu Dhaalu Atoll, Coastal Erosion

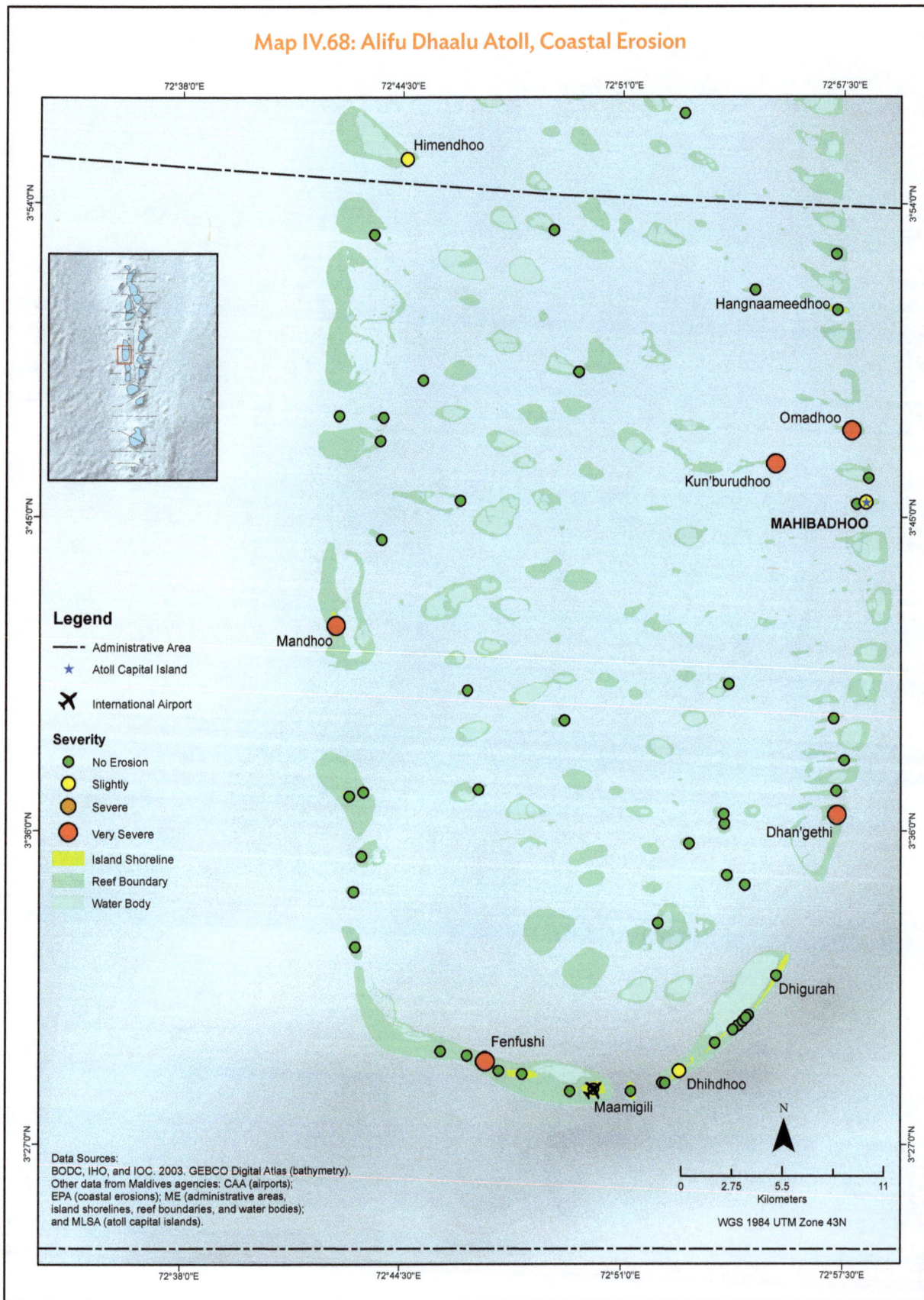

Himendhoo

Hangnaameedhoo

Omadhoo

Kun'burudhoo

MAHIBADHOO

Mandhoo

Legend

— Administrative Area

★ Atoll Capital Island

✈ International Airport

Severity

● No Erosion
● Slightly
● Severe
● Very Severe
Island Shoreline
Reef Boundary
Water Body

Dhan'gethi

Dhigurah

Fenfushi

Dhihdhoo

Maamigili

Data Sources:
BODC, IHO, and IOC. 2003. GEBCO Digital Atlas (bathymetry).
Other data from Maldives agencies: CAA (airports);
EPA (coastal erosions); ME (administrative areas,
island shorelines, reef boundaries, and water bodies);
and MLSA (atoll capital islands).

N

0   2.75   5.5   11
Kilometers

WGS 1984 UTM Zone 43N

Map IV.69: Baa Atoll, Coastal Erosion

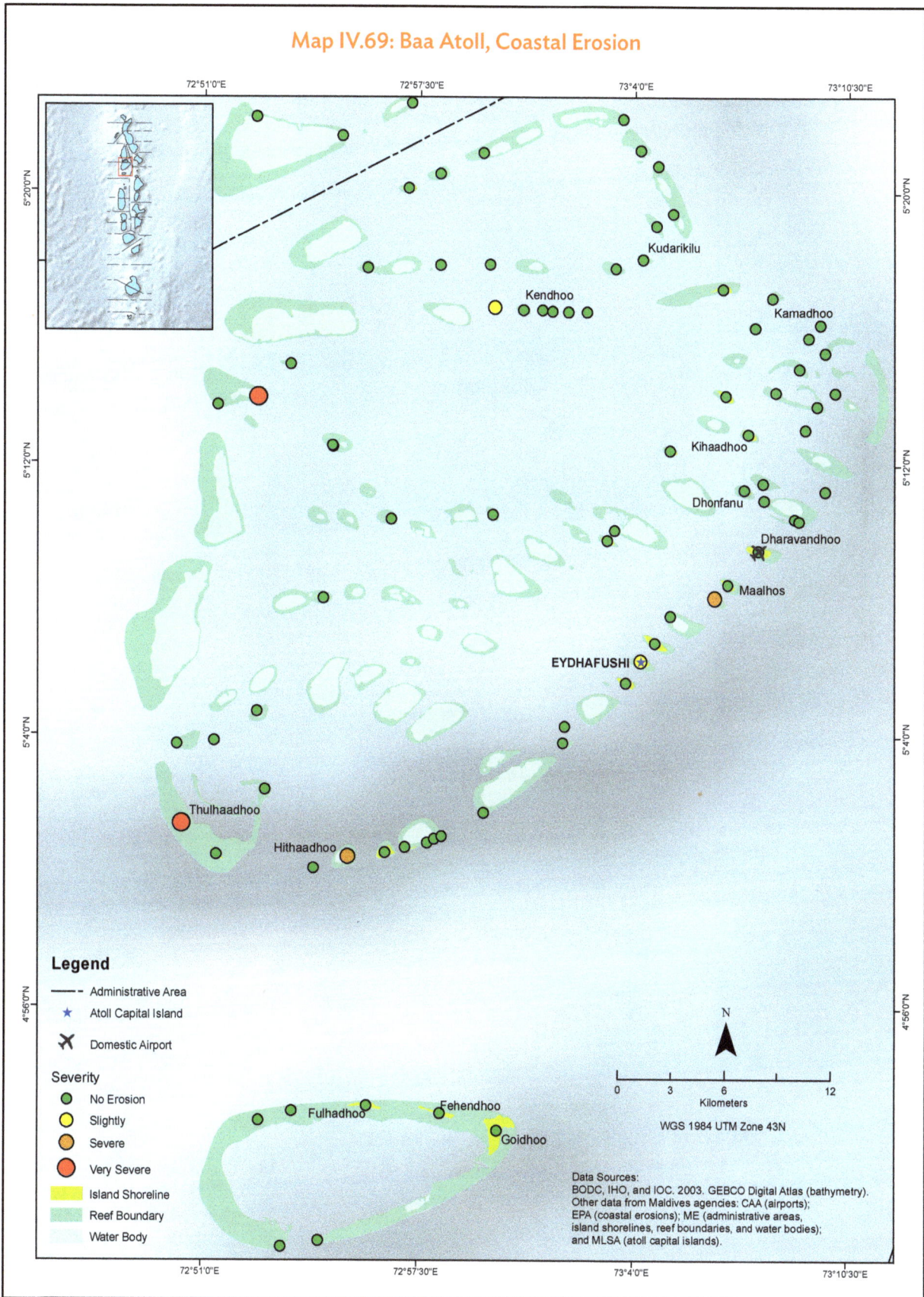

Legend

— — Administrative Area

★ Atoll Capital Island

✈ Domestic Airport

Severity

● No Erosion
● Slightly
● Severe
● Very Severe
▮ Island Shoreline
▮ Reef Boundary
　 Water Body

N

0　3　6　12
Kilometers

WGS 1984 UTM Zone 43N

Data Sources:
BODC, IHO, and IOC. 2003. GEBCO Digital Atlas (bathymetry).
Other data from Maldives agencies: CAA (airports);
EPA (coastal erosions); ME (administrative areas,
island shorelines, reef boundaries, and water bodies);
and MLSA (atoll capital islands).

Kudarikilu, Kendhoo, Kamadhoo, Kihaadhoo, Dhonfanu, Dharavandhoo, Maalhos, EYDHAFUSHI, Thulhaadhoo, Hithaadhoo, Fulhadhoo, Fehendhoo, Goidhoo

Map IV.70: Dhaalu Atoll, Coastal Erosion

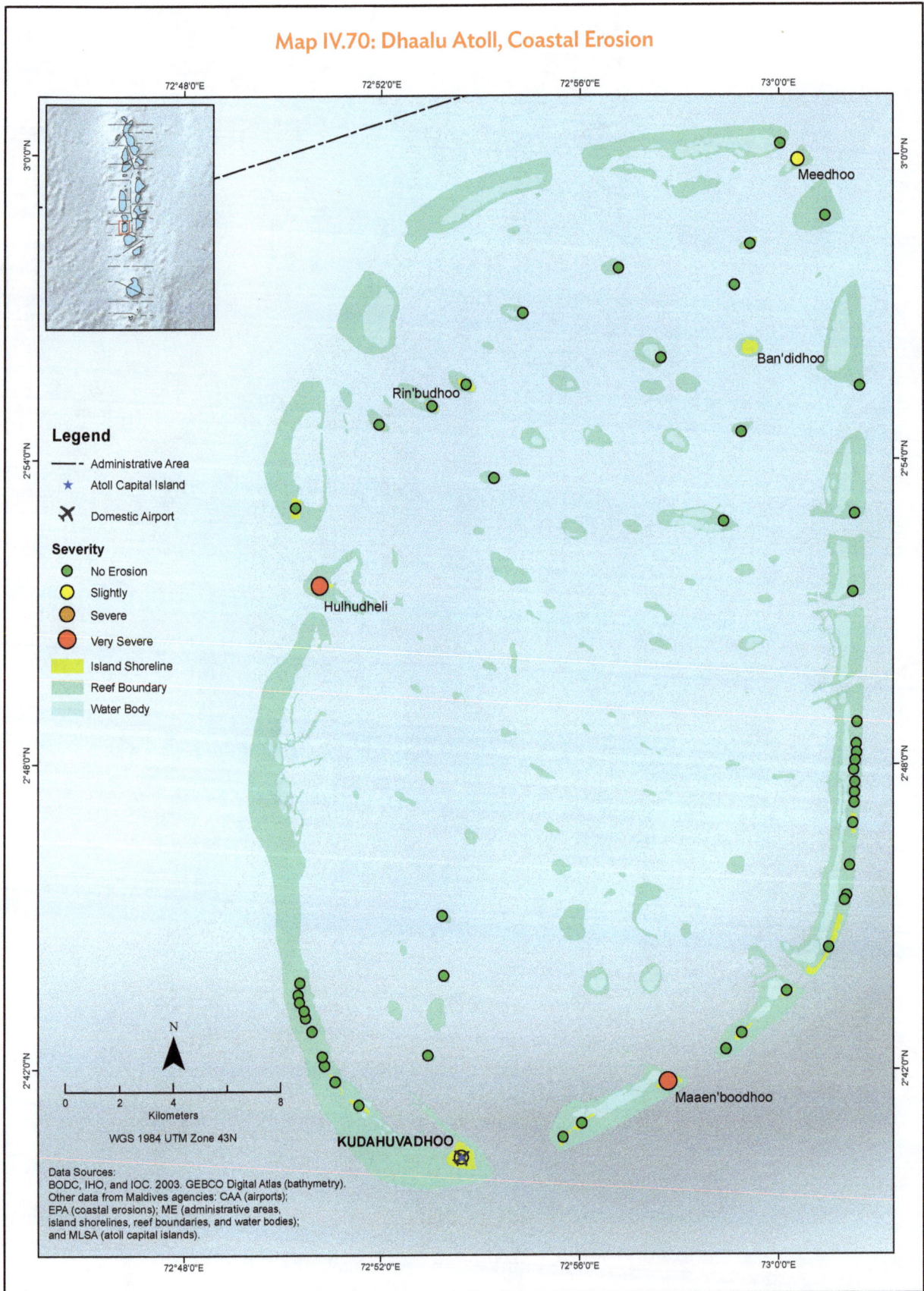

Legend

— Administrative Area

★ Atoll Capital Island

✈ Domestic Airport

Severity
- ● No Erosion
- ● Slightly
- ● Severe
- ● Very Severe
- ▩ Island Shoreline
- ▩ Reef Boundary
- ▩ Water Body

Meedhoo

Ban'didhoo

Rin'budhoo

Hulhudheli

Maaen'boodhoo

KUDAHUVADHOO

N

0   2   4   8
Kilometers

WGS 1984 UTM Zone 43N

Data Sources:
BODC, IHO, and IOC. 2003. GEBCO Digital Atlas (bathymetry).
Other data from Maldives agencies: CAA (airports);
EPA (coastal erosions); ME (administrative areas,
island shorelines, reef boundaries, and water bodies);
and MLSA (atoll capital islands).

Map IV.71: Faafu Atoll, Coastal Erosion

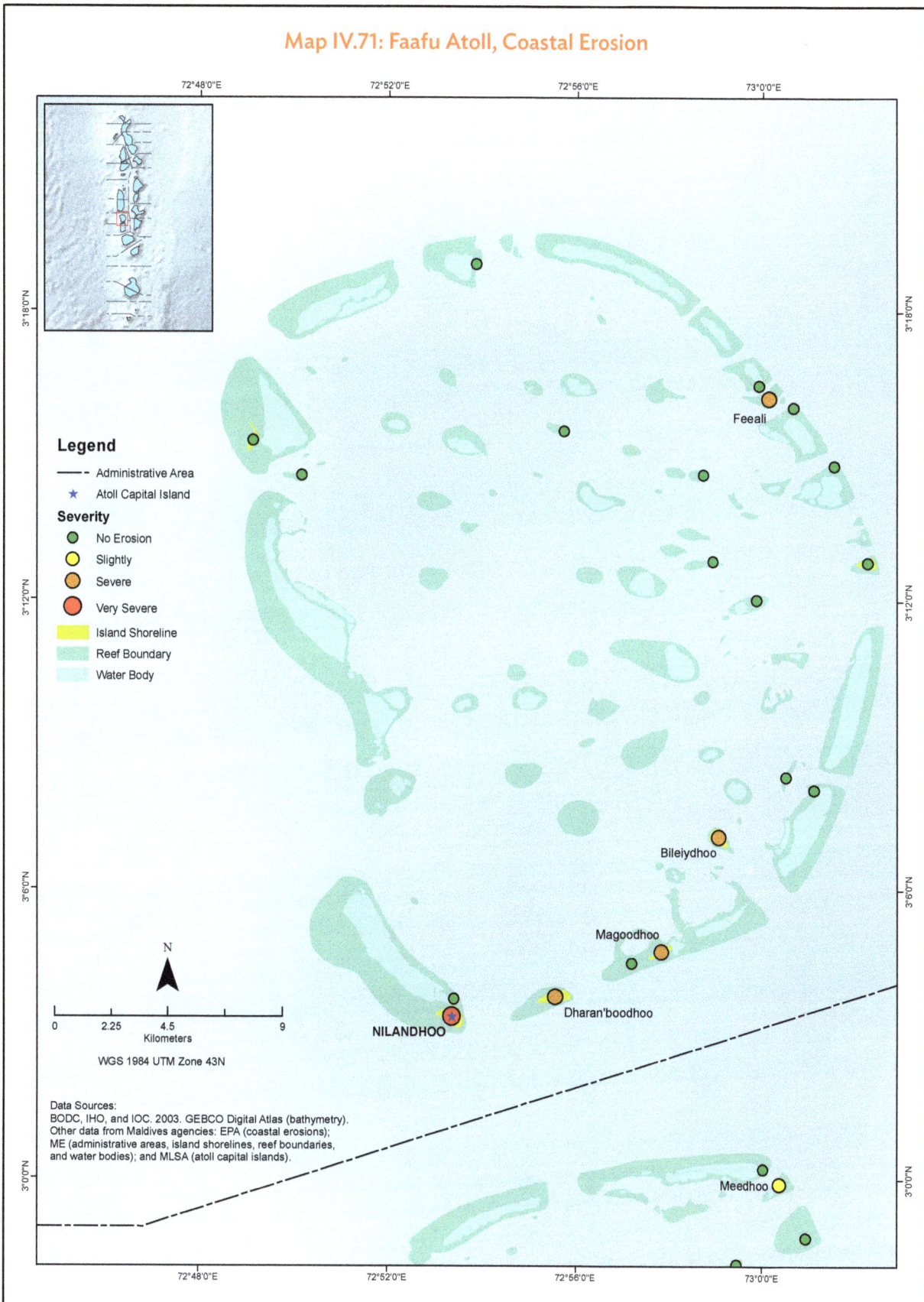

Legend

— — Administrative Area

★ Atoll Capital Island

Severity
- No Erosion
- Slightly
- Severe
- Very Severe
- Island Shoreline
- Reef Boundary
- Water Body

N

0   2.25   4.5   9
Kilometers

WGS 1984 UTM Zone 43N

Data Sources:
BODC, IHO, and IOC. 2003. GEBCO Digital Atlas (bathymetry).
Other data from Maldives agencies: EPA (coastal erosions);
ME (administrative areas, island shorelines, reef boundaries,
and water bodies); and MLSA (atoll capital islands).

Feeali

Bileiydhoo

Magoodhoo

Dharan'boodhoo

NILANDHOO

Meedhoo

Map IV.72: Gaafu Alifu Atoll, Coastal Erosion

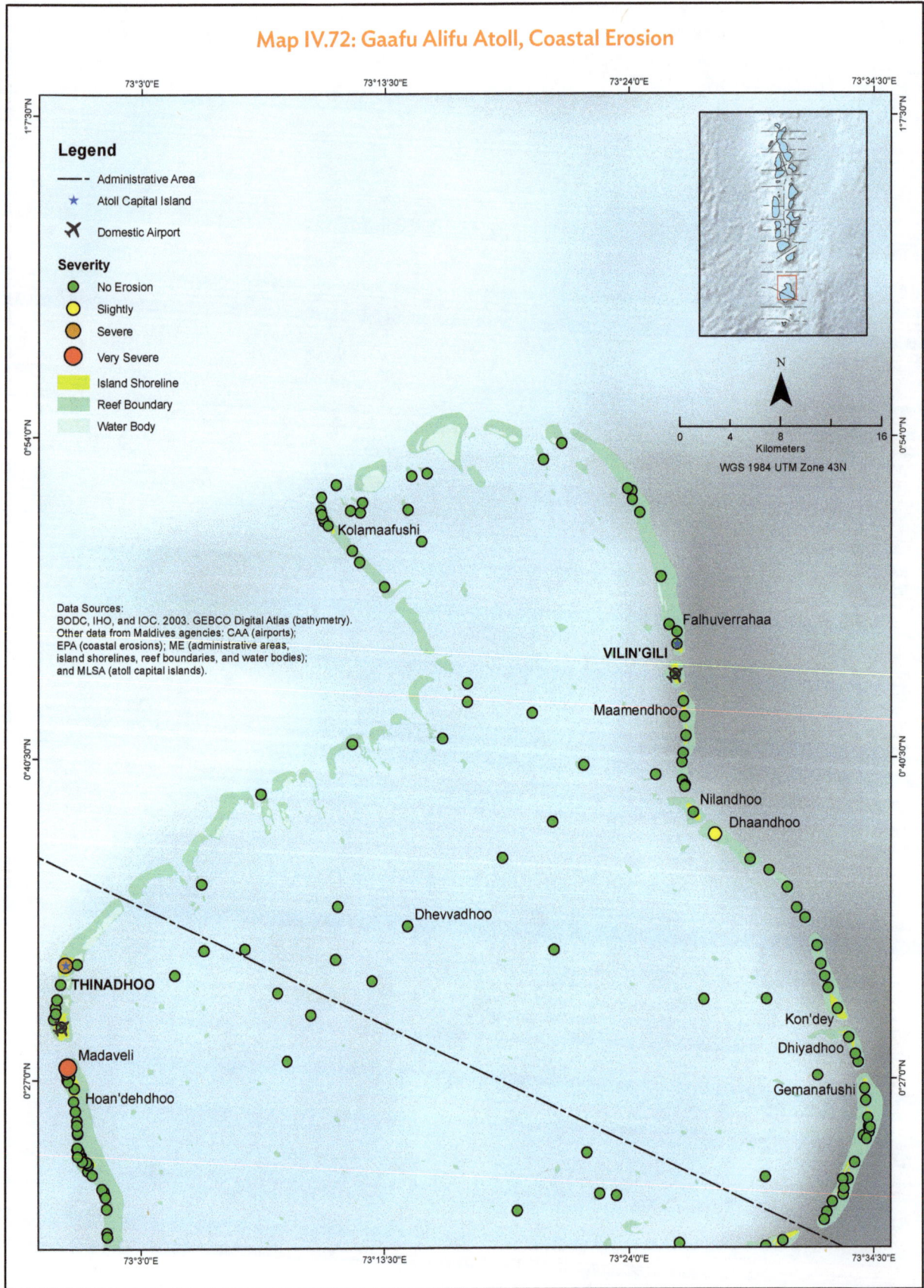

Legend

— - — Administrative Area

★ Atoll Capital Island

✈ Domestic Airport

Severity

● No Erosion
● Slightly
● Severe
● Very Severe

▮ Island Shoreline
▮ Reef Boundary
▮ Water Body

Data Sources:
BODC, IHO, and IOC. 2003. GEBCO Digital Atlas (bathymetry).
Other data from Maldives agencies: CAA (airports);
EPA (coastal erosions); ME (administrative areas,
island shorelines, reef boundaries, and water bodies);
and MLSA (atoll capital islands).

N

0   4   8   16
Kilometers
WGS 1984 UTM Zone 43N

Kolamaafushi

Falhuverrahaa

VILIN'GILI

Maamendhoo

Nilandhoo

Dhaandhoo

Dhevvadhoo

THINADHOO

Kon'dey

Dhiyadhoo

Madaveli

Gemanafushi

Hoan'dehdhoo

Map IV.73: Gaafu Dhaalu Atoll, Coastal Erosion

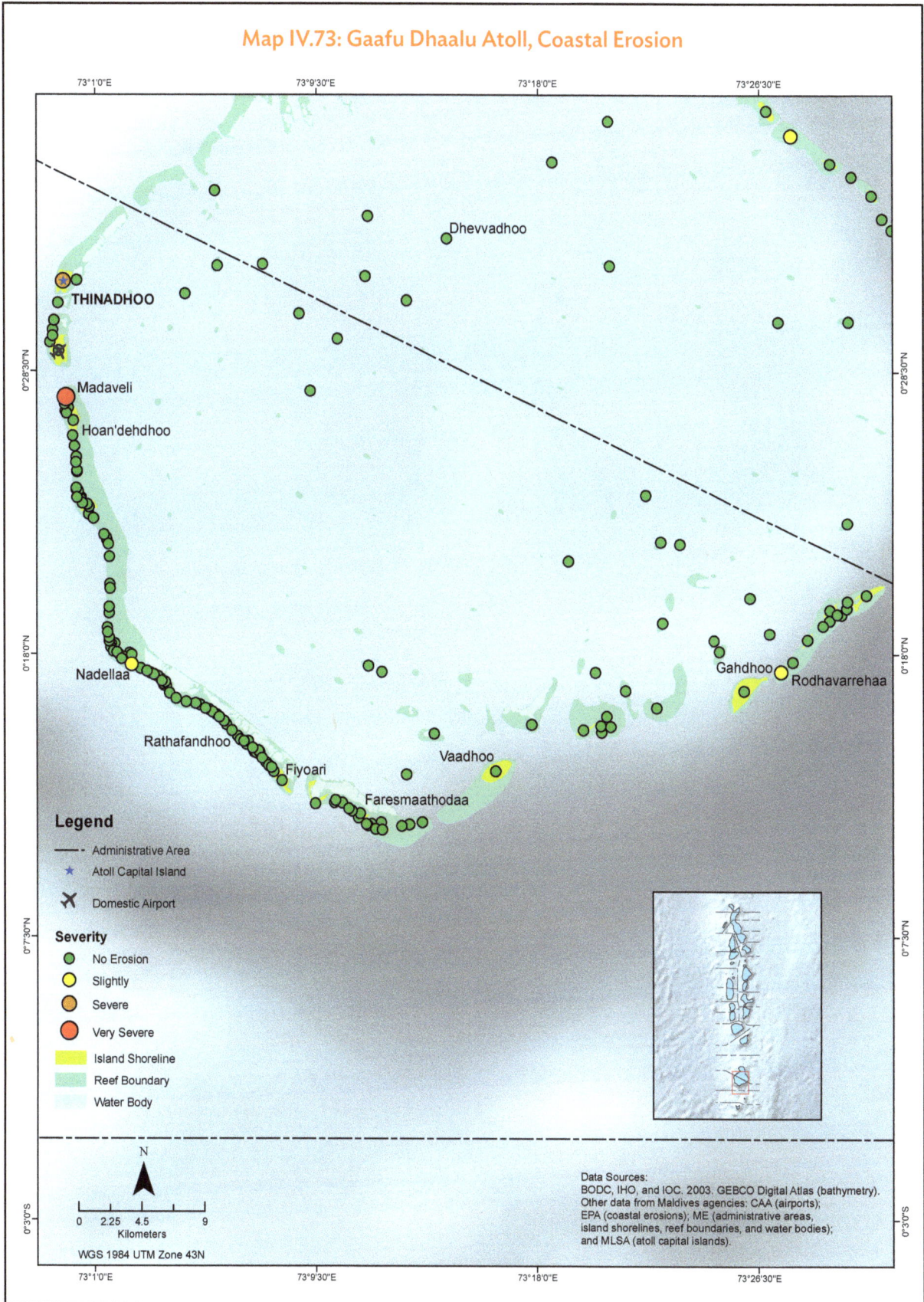

Dhevvadhoo

THINADHOO

Madaveli

Hoan'dehdhoo

Nadellaa

Rathafandhoo

Fiyoari

Faresmaathodaa

Vaadhoo

Gahdhoo

Rodhavarrehaa

Legend

— · — Administrative Area

★ Atoll Capital Island

✈ Domestic Airport

Severity

🟢 No Erosion

🟡 Slightly

🟠 Severe

🔴 Very Severe

🟨 Island Shoreline

🟩 Reef Boundary

⬜ Water Body

N

0   2.25   4.5   9
Kilometers
WGS 1984 UTM Zone 43N

Data Sources:
BODC, IHO, and IOC. 2003. GEBCO Digital Atlas (bathymetry).
Other data from Maldives agencies: CAA (airports);
EPA (coastal erosions); ME (administrative areas,
island shorelines, reef boundaries, and water bodies);
and MLSA (atoll capital islands).

Map IV.74: Gnaviyani Atoll, Coastal Erosion

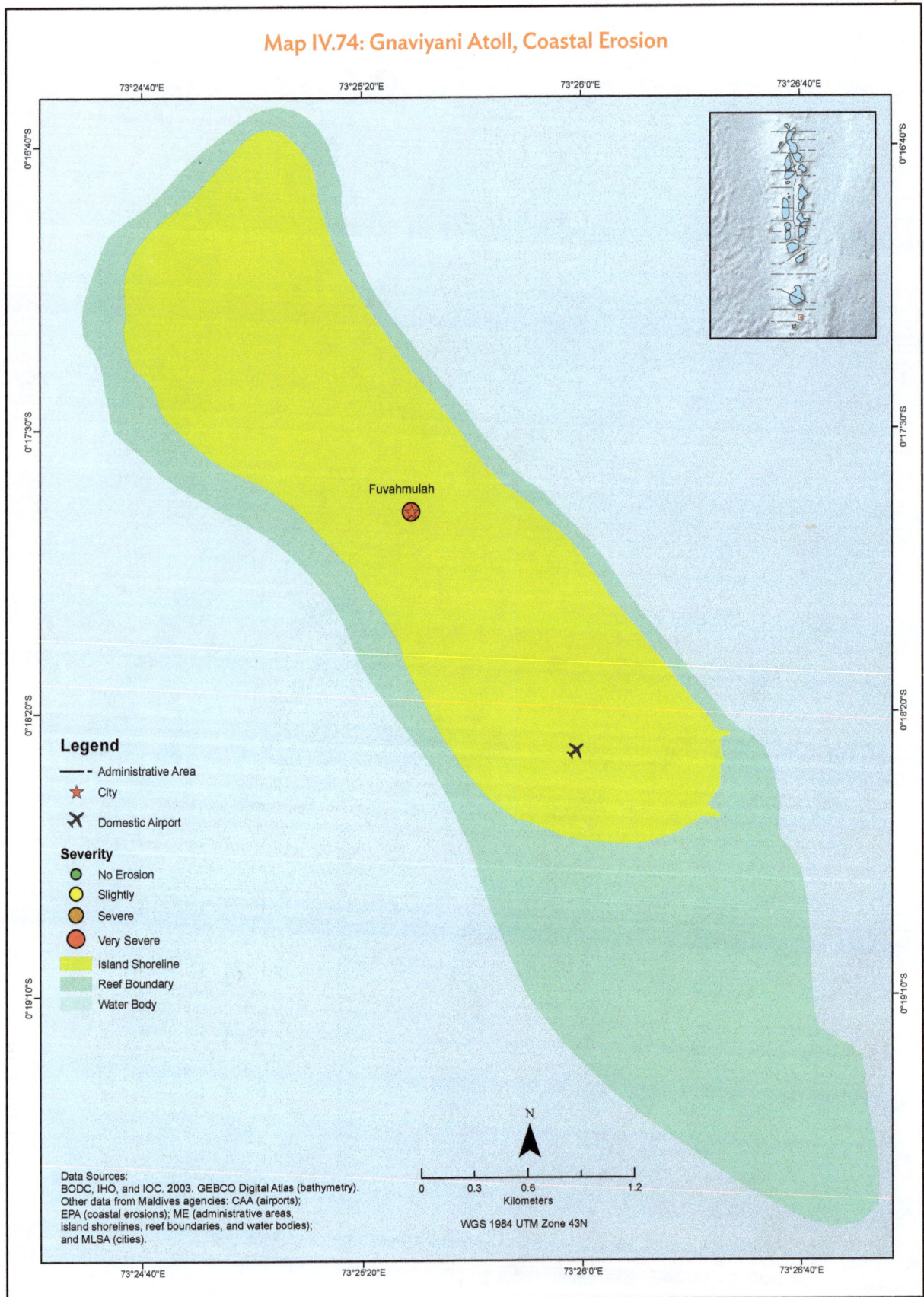

Fuvahmulah

Legend

— – Administrative Area

⭐ City

✈ Domestic Airport

Severity

🟢 No Erosion

🟡 Slightly

🟠 Severe

🔴 Very Severe

🟨 Island Shoreline

🟩 Reef Boundary

⬜ Water Body

Data Sources:
BODC, IHO, and IOC. 2003. GEBCO Digital Atlas (bathymetry).
Other data from Maldives agencies: CAA (airports);
EPA (coastal erosions); ME (administrative areas,
island shorelines, reef boundaries, and water bodies);
and MLSA (cities).

N

0   0.3   0.6   1.2
Kilometers

WGS 1984 UTM Zone 43N

Map IV.75: Haa Alifu Atoll, Coastal Erosion

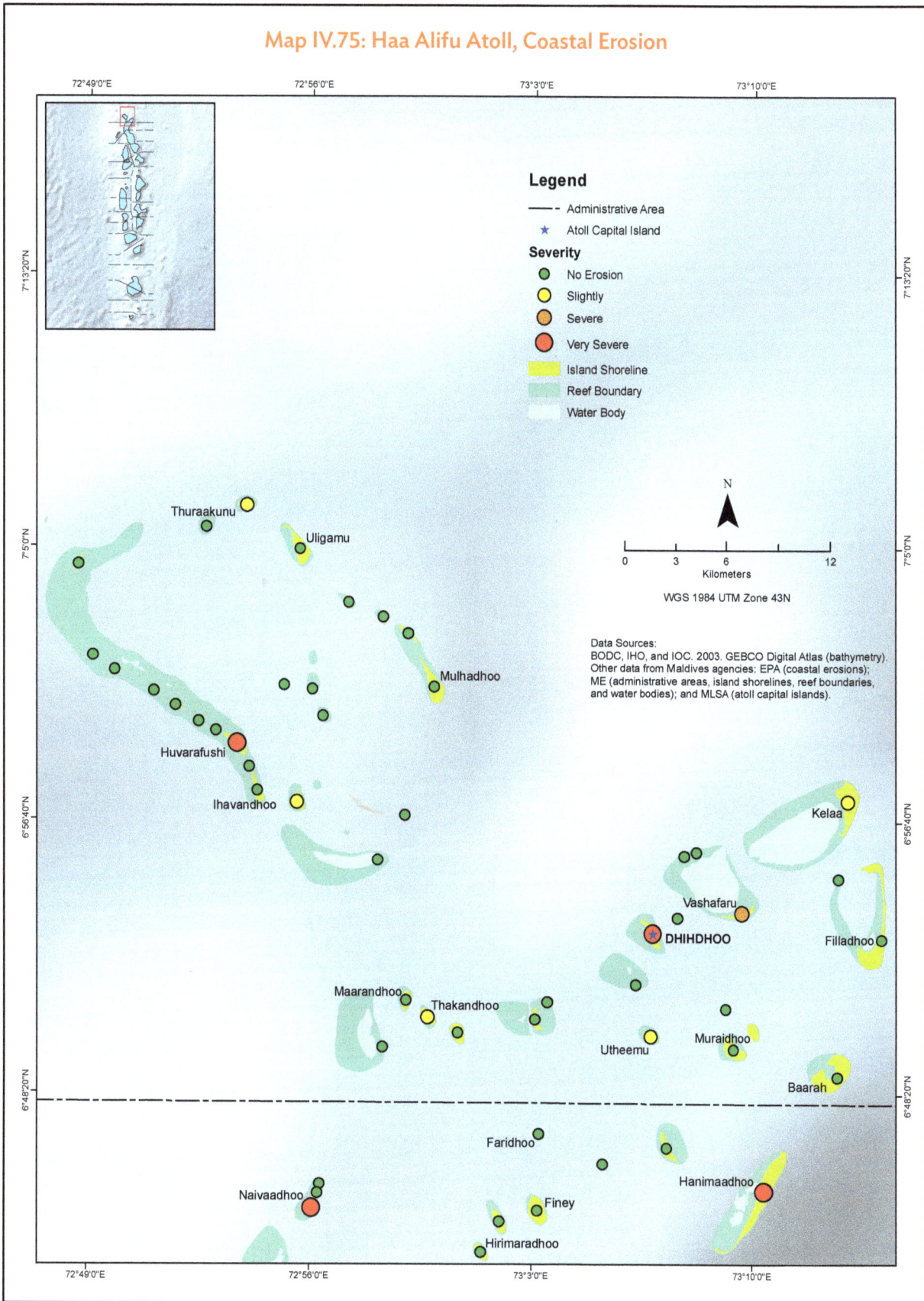

Legend

- - – Administrative Area
- ★ Atoll Capital Island

Severity
- ● No Erosion
- ● Slightly
- ● Severe
- ● Very Severe

- Island Shoreline
- Reef Boundary
- Water Body

N

| 0 | 3 | 6 | 12 |
Kilometers

WGS 1984 UTM Zone 43N

Data Sources:
BODC, IHO, and IOC. 2003. GEBCO Digital Atlas (bathymetry).
Other data from Maldives agencies: EPA (coastal erosions);
ME (administrative areas, island shorelines, reef boundaries,
and water bodies); and MLSA (atoll capital islands).

Thuraakunu
Uligamu
Mulhadhoo
Huvarafushi
Ihavandhoo
Kelaa
Vashafaru
DHIHDHOO
Filladhoo
Maarandhoo
Thakandhoo
Utheemu
Muraidhoo
Baarah
Faridhoo
Hanimaadhoo
Naivaadhoo
Finey
Hirimaradhoo

Map IV.76: Haa Dhaalu Atoll, Coastal Erosion

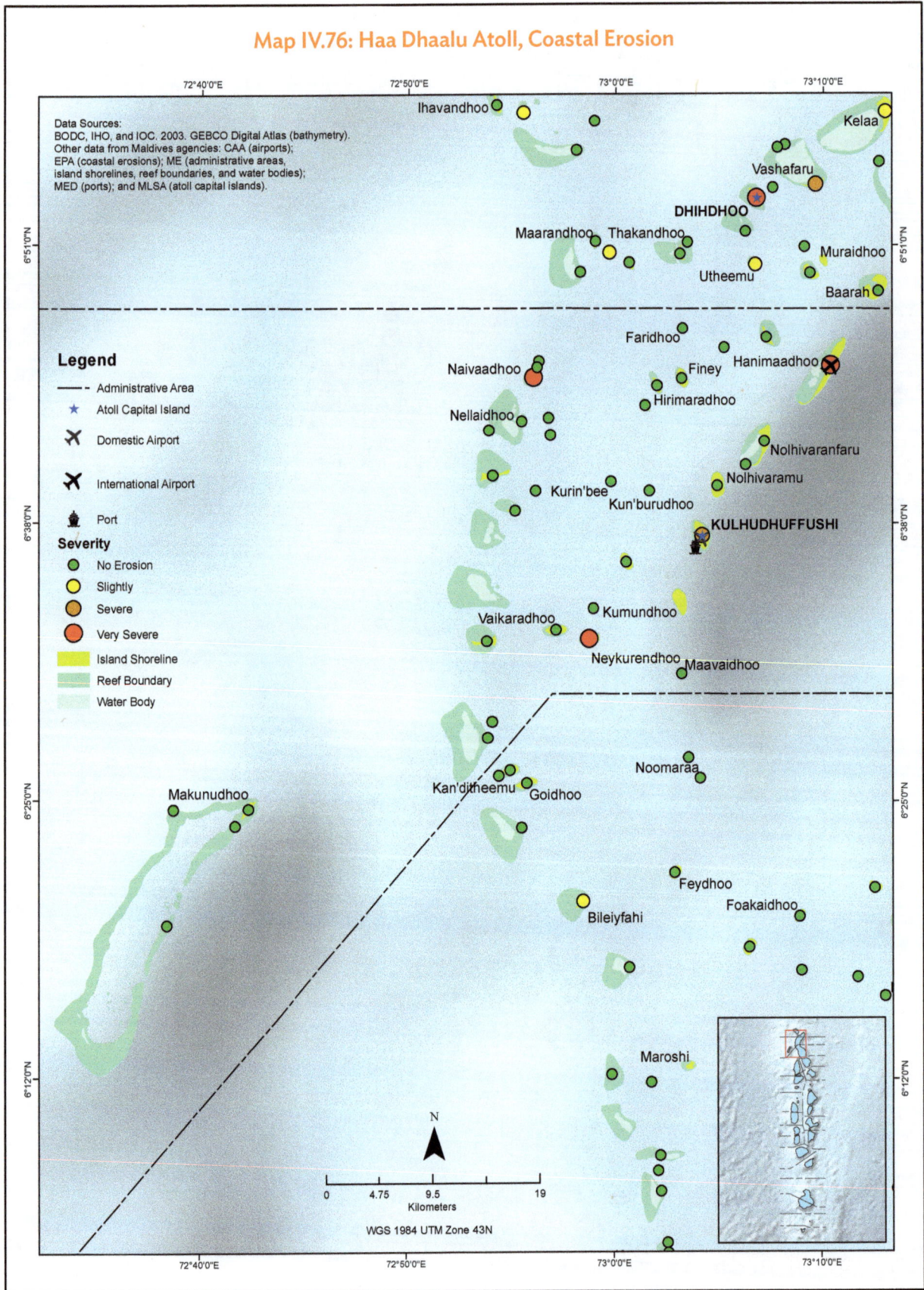

Data Sources:
BODC, IHO, and IOC. 2003. GEBCO Digital Atlas (bathymetry).
Other data from Maldives agencies: CAA (airports);
EPA (coastal erosions); ME (administrative areas,
island shorelines, reef boundaries, and water bodies);
MED (ports); and MLSA (atoll capital islands).

Ihavandhoo

Kelaa

Vashafaru

DHIHDHOO

Maarandhoo   Thakandhoo

Muraidhoo

Utheemu

Baarah

Faridhoo

Finey   Hanimaadhoo

Naivaadhoo

Hirimaradhoo

Nellaidhoo

Nolhivaranfaru

Nolhivaramu

Kurin'bee

Kun'burudhoo

KULHUDHUFFUSHI

Kumundhoo

Vaikaradhoo

Neykurendhoo   Maavaidhoo

Noomaraa

Kan'ditheemu   Goidhoo

Makunudhoo

Feydhoo

Foakaidhoo

Bileiyfahi

Maroshi

Legend

- – – Administrative Area
- ★ Atoll Capital Island
- ✈ Domestic Airport
- ✈ International Airport
- ⚓ Port

Severity

- 🟢 No Erosion
- 🟡 Slightly
- 🟠 Severe
- 🔴 Very Severe
- 🟨 Island Shoreline
- 🟩 Reef Boundary
- 🟩 Water Body

N

0   4.75   9.5   19
Kilometers

WGS 1984 UTM Zone 43N

Map IV.77: Laamu Atoll, Coastal Erosion

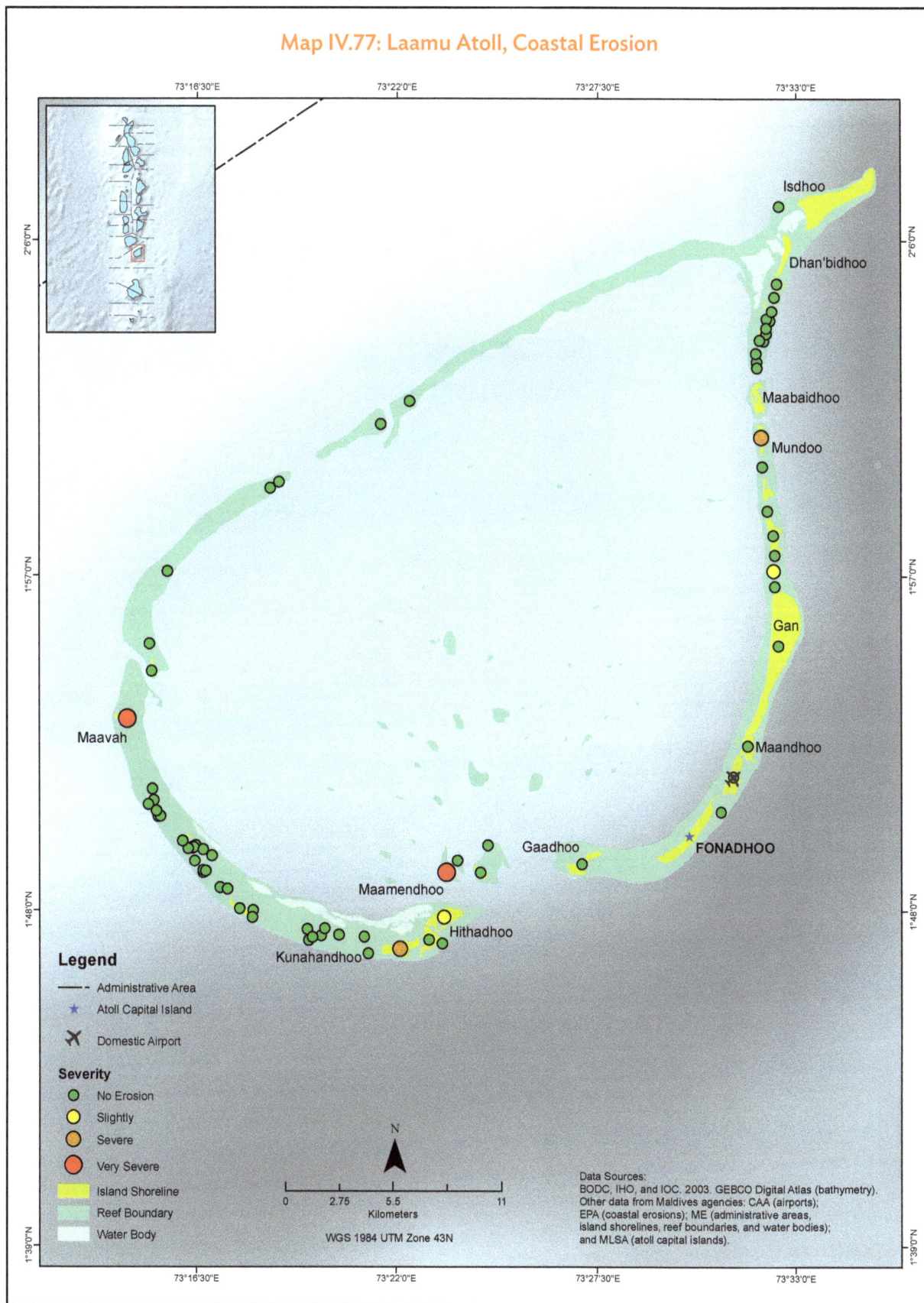

Isdhoo

Dhan'bidhoo

Maabaidhoo

Mundoo

Gan

Maandhoo

FONADHOO

Maavah

Gaadhoo

Maamendhoo

Hithadhoo

Kunahandhoo

Legend

— - Administrative Area

★ Atoll Capital Island

✕ Domestic Airport

Severity

◯ No Erosion

◯ Slightly

◯ Severe

◯ Very Severe

▮ Island Shoreline

▮ Reef Boundary

▮ Water Body

N

0   2.75   5.5   11
Kilometers

WGS 1984 UTM Zone 43N

Data Sources:
BODC, IHO, and IOC. 2003. GEBCO Digital Atlas (bathymetry).
Other data from Maldives agencies: CAA (airports);
EPA (coastal erosions); ME (administrative areas,
island shorelines, reef boundaries, and water bodies);
and MLSA (atoll capital islands).

Map IV.78: Lhaviyani Atoll, Coastal Erosion

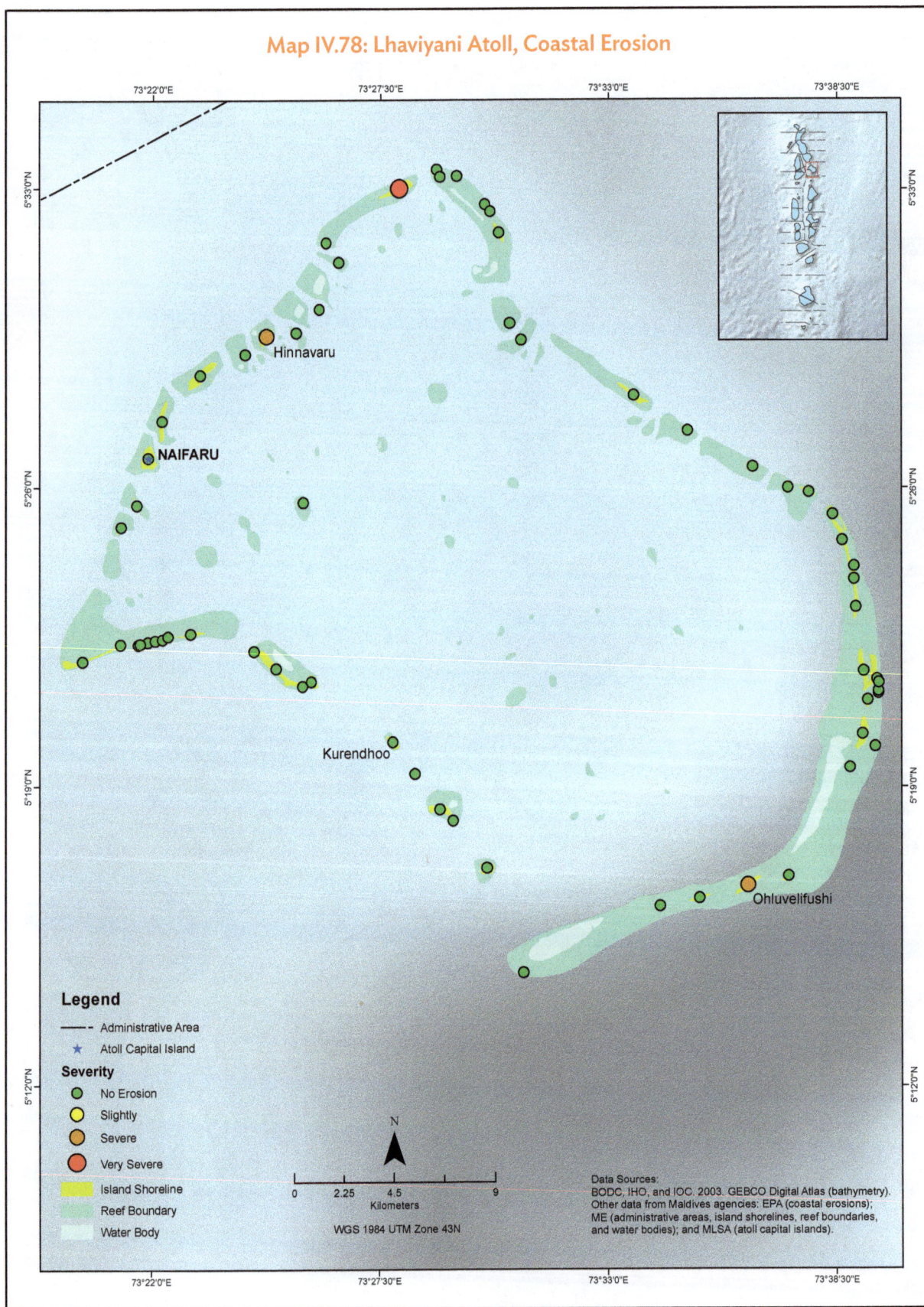

Hinnavaru

NAIFARU

Kurendhoo

Ohluvelifushi

Legend

– – – Administrative Area

★ Atoll Capital Island

Severity

- ● No Erosion
- ● Slightly
- ● Severe
- ● Very Severe
- ▬ Island Shoreline
- ▬ Reef Boundary
- ▬ Water Body

N

0   2.25   4.5   9
Kilometers

WGS 1984 UTM Zone 43N

Data Sources:
BODC, IHO, and IOC. 2003. GEBCO Digital Atlas (bathymetry).
Other data from Maldives agencies: EPA (coastal erosions);
ME (administrative areas, island shorelines, reef boundaries,
and water bodies); and MLSA (atoll capital islands).

Map IV.79: Meemu Atoll, Coastal Erosion

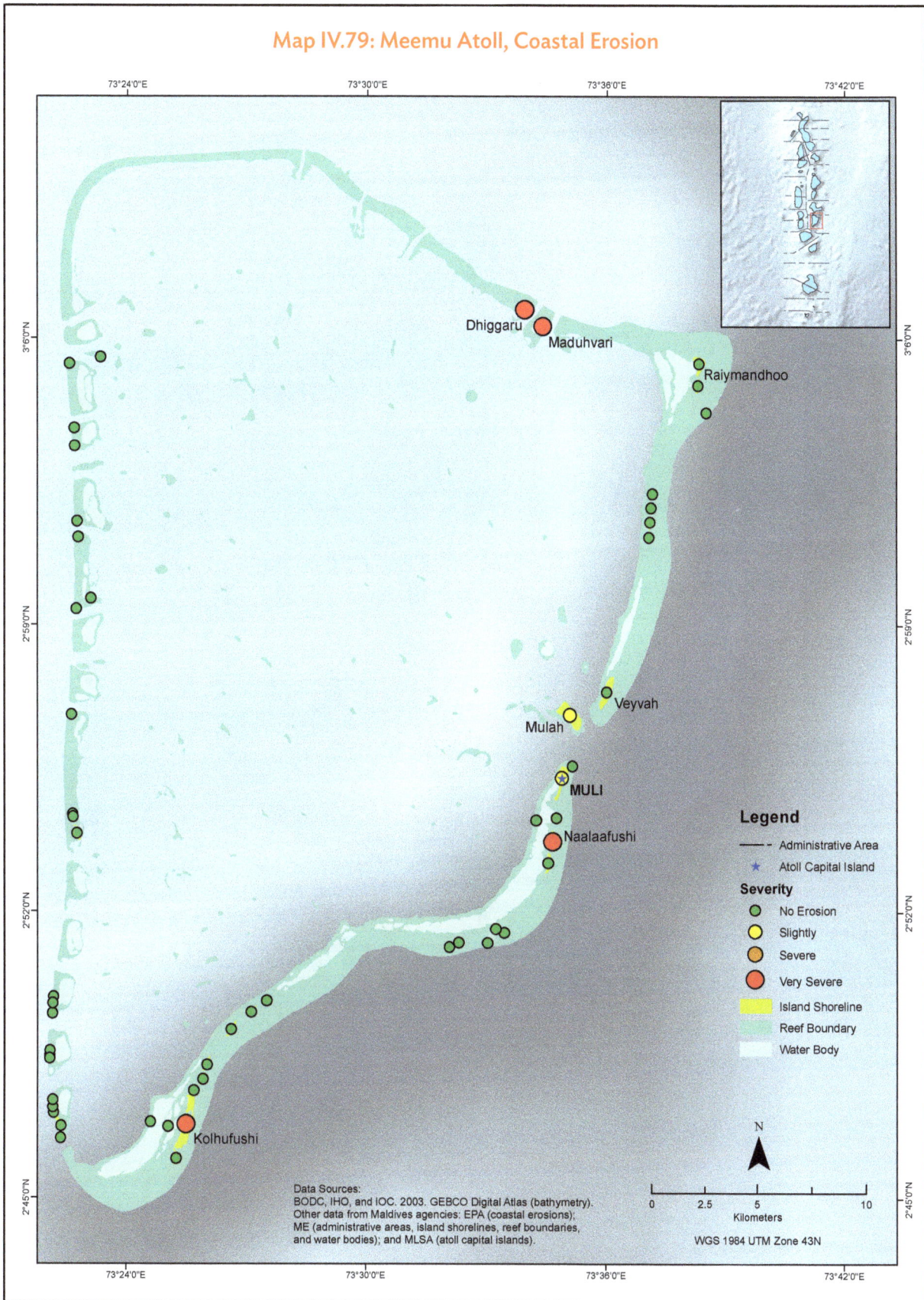

Dhiggaru
Maduhvari
Raiymandhoo
Veyvah
Mulah
MULI
Naalaafushi
Kolhufushi

Legend

— Administrative Area
★ Atoll Capital Island

Severity
● No Erosion
● Slightly
● Severe
● Very Severe
▬ Island Shoreline
Reef Boundary
Water Body

N

Data Sources:
BODC, IHO, and IOC. 2003. GEBCO Digital Atlas (bathymetry).
Other data from Maldives agencies: EPA (coastal erosions);
ME (administrative areas, island shorelines, reef boundaries,
and water bodies); and MLSA (atoll capital islands).

0   2.5   5   10
Kilometers

WGS 1984 UTM Zone 43N

Map IV.80: Noonu Atoll, Coastal Erosion

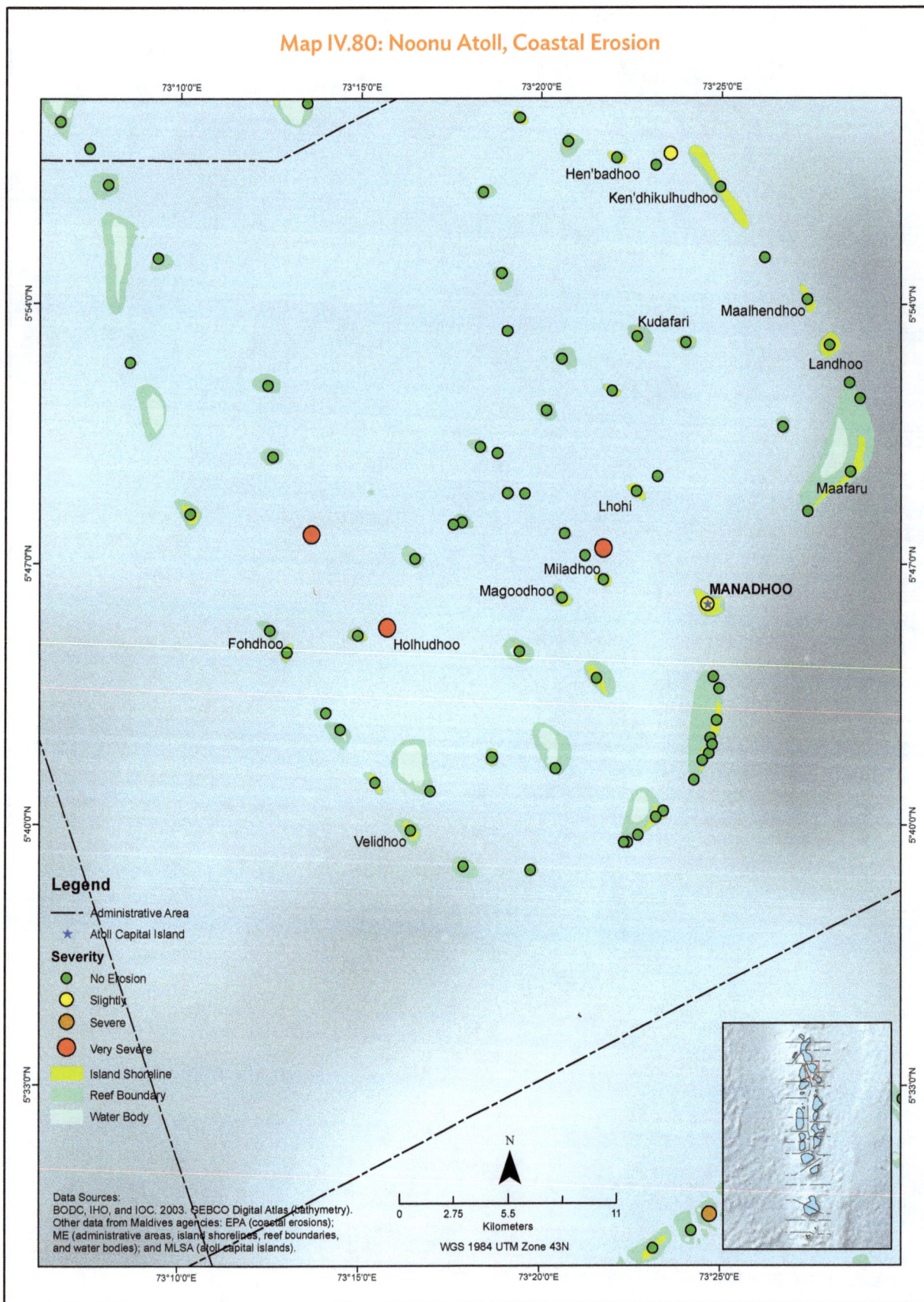

Hen'badhoo

Ken'dhikulhudhoo

Maalhendhoo

Kudafari

Landhoo

Maafaru

Lhohi

Miladhoo

Magoodhoo

MANADHOO

Fohdhoo

Holhudhoo

Velidhoo

Legend

- - - Administrative Area

★ Atoll Capital Island

Severity

- 🟢 No Erosion
- 🟡 Slightly
- 🟠 Severe
- 🔴 Very Severe
- 🟨 Island Shoreline
- 🟩 Reef Boundary
- ⬜ Water Body

Data Sources:
BODC, IHO, and IOC. 2003. GEBCO Digital Atlas (bathymetry).
Other data from Maldives agencies: EPA (coastal erosions);
ME (administrative areas, island shorelines, reef boundaries,
and water bodies); and MLSA (atoll capital islands).

N

0   2.75   5.5   11
Kilometers

WGS 1984 UTM Zone 43N

Map IV.81: North Malé Atoll, Coastal Erosion

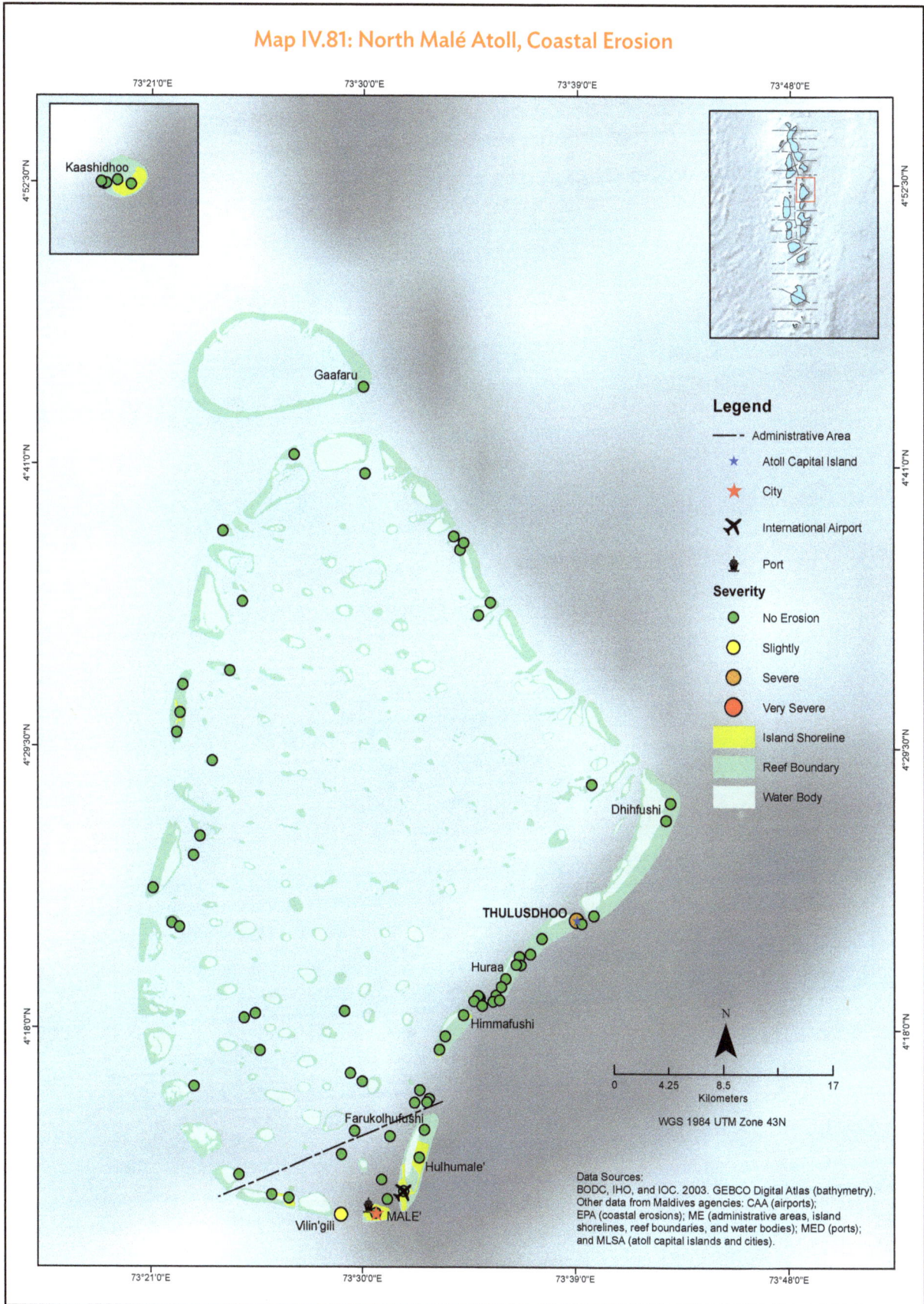

Kaashidhoo

Gaafaru

Dhihfushi

THULUSDHOO

Huraa

Himmafushi

Farukolhufushi

Hulhumale'

Vilin'gili   MALE'

Legend

- – · – Administrative Area
- ★ Atoll Capital Island
- ★ City
- ✈ International Airport
- ⚓ Port

Severity
- ● No Erosion
- ● Slightly
- ● Severe
- ● Very Severe
- ▉ Island Shoreline
- ▉ Reef Boundary
- ▢ Water Body

N

| 0 | 4.25 | 8.5 | | 17 |

Kilometers

WGS 1984 UTM Zone 43N

Data Sources:
BODC, IHO, and IOC. 2003. GEBCO Digital Atlas (bathymetry).
Other data from Maldives agencies: CAA (airports);
EPA (coastal erosions); ME (administrative areas, island
shorelines, reef boundaries, and water bodies); MED (ports);
and MLSA (atoll capital islands and cities).

Map IV.82: Raa Atoll, Coastal Erosion

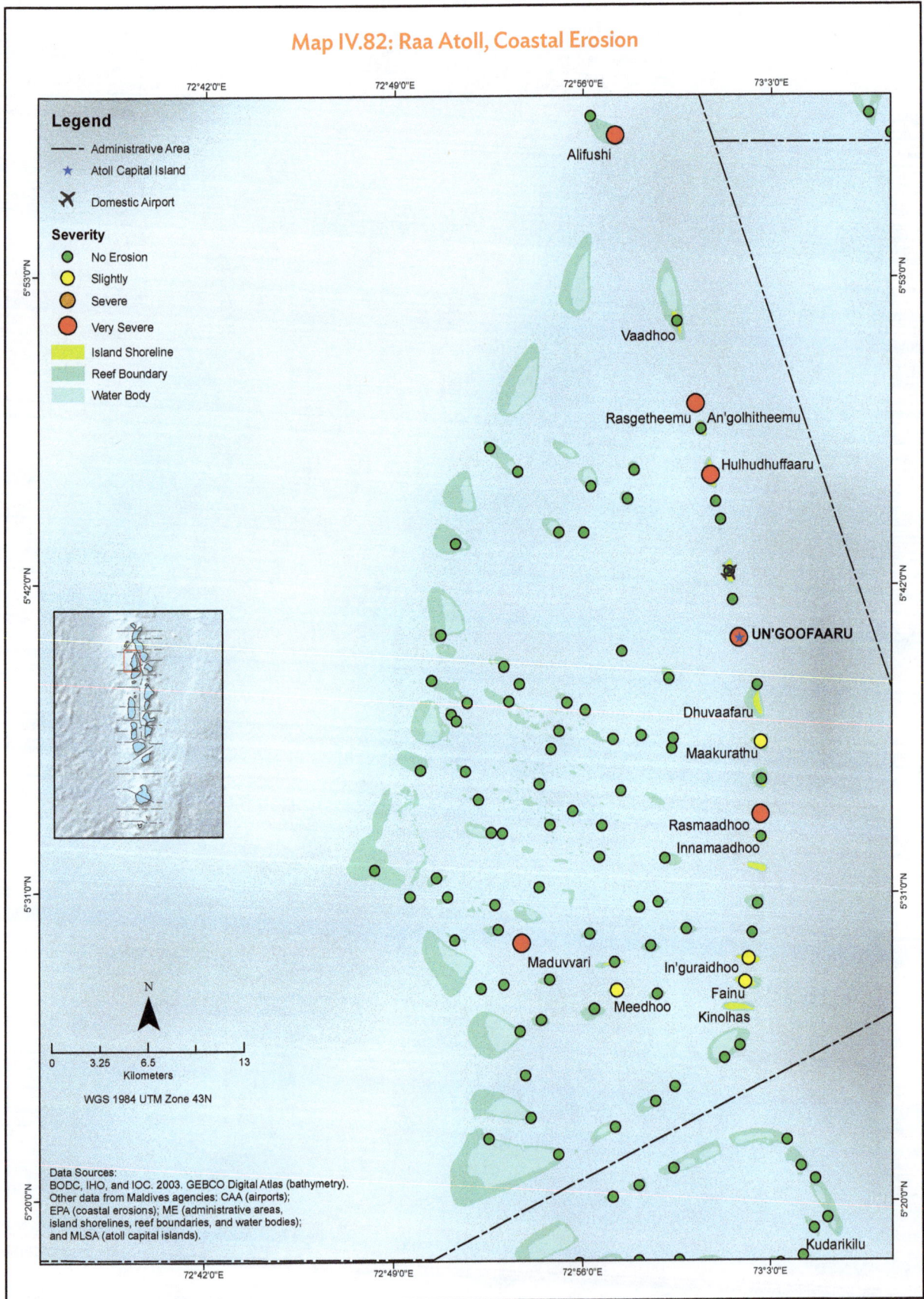

Legend

– – – Administrative Area

★ Atoll Capital Island

✈ Domestic Airport

Severity
- 🟢 No Erosion
- 🟡 Slightly
- 🟠 Severe
- 🔴 Very Severe
- 🟨 Island Shoreline
- 🟩 Reef Boundary
- 🟦 Water Body

Alifushi

Vaadhoo

Rasgetheemu An'golhitheemu

Hulhudhuffaaru

UN'GOOFAARU

Dhuvaafaru

Maakurathu

Rasmaadhoo

Innamaadhoo

Maduvvari

In'guraidhoo

Fainu

Meedhoo

Kinolhas

Kudarikilu

0   3.25   6.5   13
Kilometers

WGS 1984 UTM Zone 43N

Data Sources:
BODC, IHO, and IOC. 2003. GEBCO Digital Atlas (bathymetry).
Other data from Maldives agencies: CAA (airports);
EPA (coastal erosions); ME (administrative areas,
island shorelines, reef boundaries, and water bodies);
and MLSA (atoll capital islands).

Map IV.83: Shaviyani Atoll, Coastal Erosion

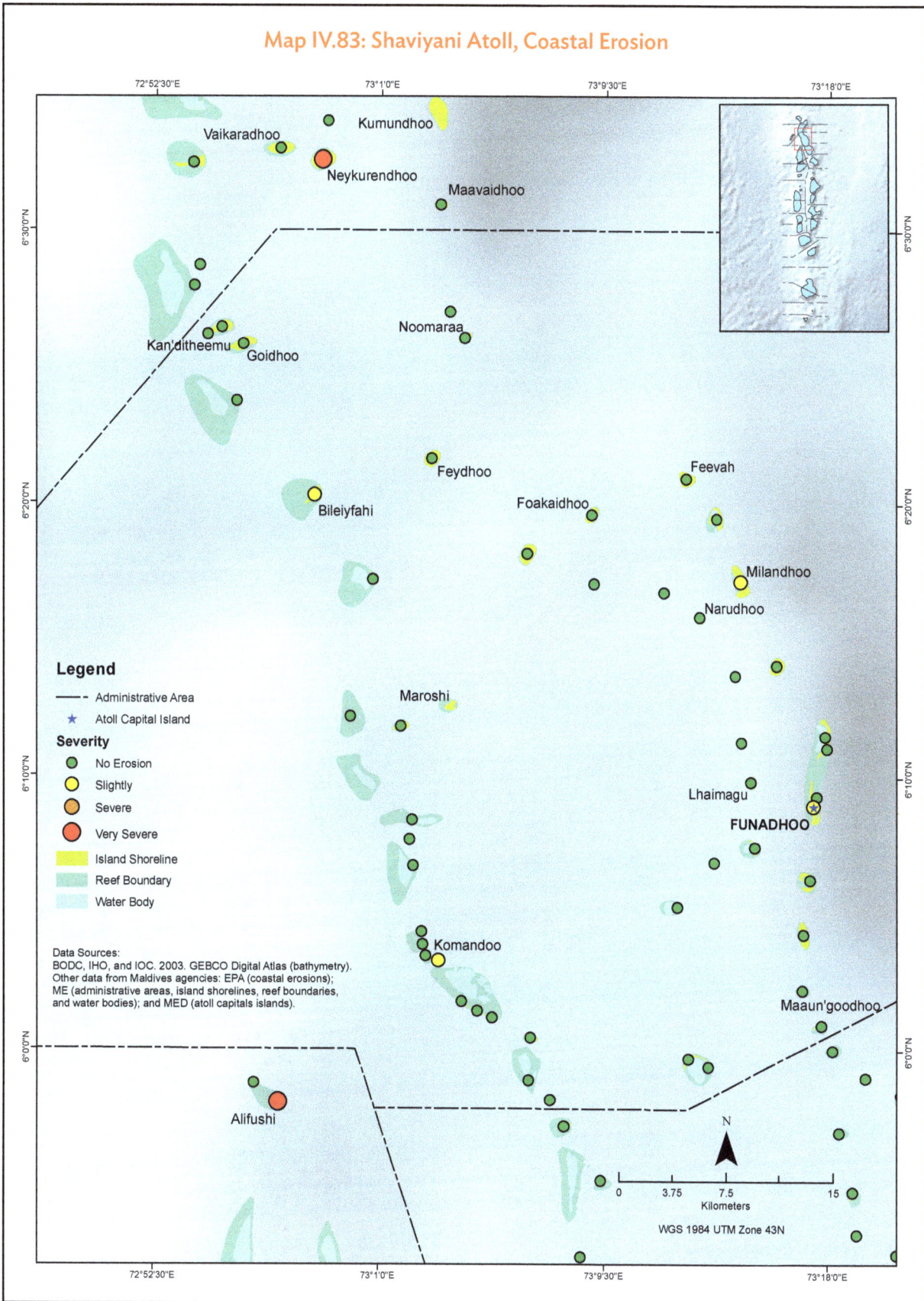

72°52'30"E   73°1'0"E   73°9'30"E   73°18'0"E

Kumundhoo

Vaikaradhoo

Neykurendhoo

Maavaidhoo

6°30'0"N

Kan'ditheemu

Goidhoo

Noomaraa

6°20'0"N

Feydhoo

Feevah

Bileiyfahi

Foakaidhoo

Milandhoo

Narudhoo

Legend

— · — Administrative Area

★ Atoll Capital Island

Severity

● No Erosion

● Slightly

● Severe

● Very Severe

Maroshi

6°10'0"N

Lhaimagu

FUNADHOO

Island Shoreline

Reef Boundary

Water Body

Komandoo

Maaun'goodhoo

Data Sources:
BODC, IHO, and IOC. 2003. GEBCO Digital Atlas (bathymetry).
Other data from Maldives agencies: EPA (coastal erosions);
ME (administrative areas, island shorelines, reef boundaries,
and water bodies); and MED (atoll capitals islands).

6°0'0"N

Alifushi

N

0   3.75   7.5   15
Kilometers

WGS 1984 UTM Zone 43N

72°52'30"E   73°1'0"E   73°9'30"E   73°18'0"E

Map IV.84: South Malé Atoll, Coastal Erosion

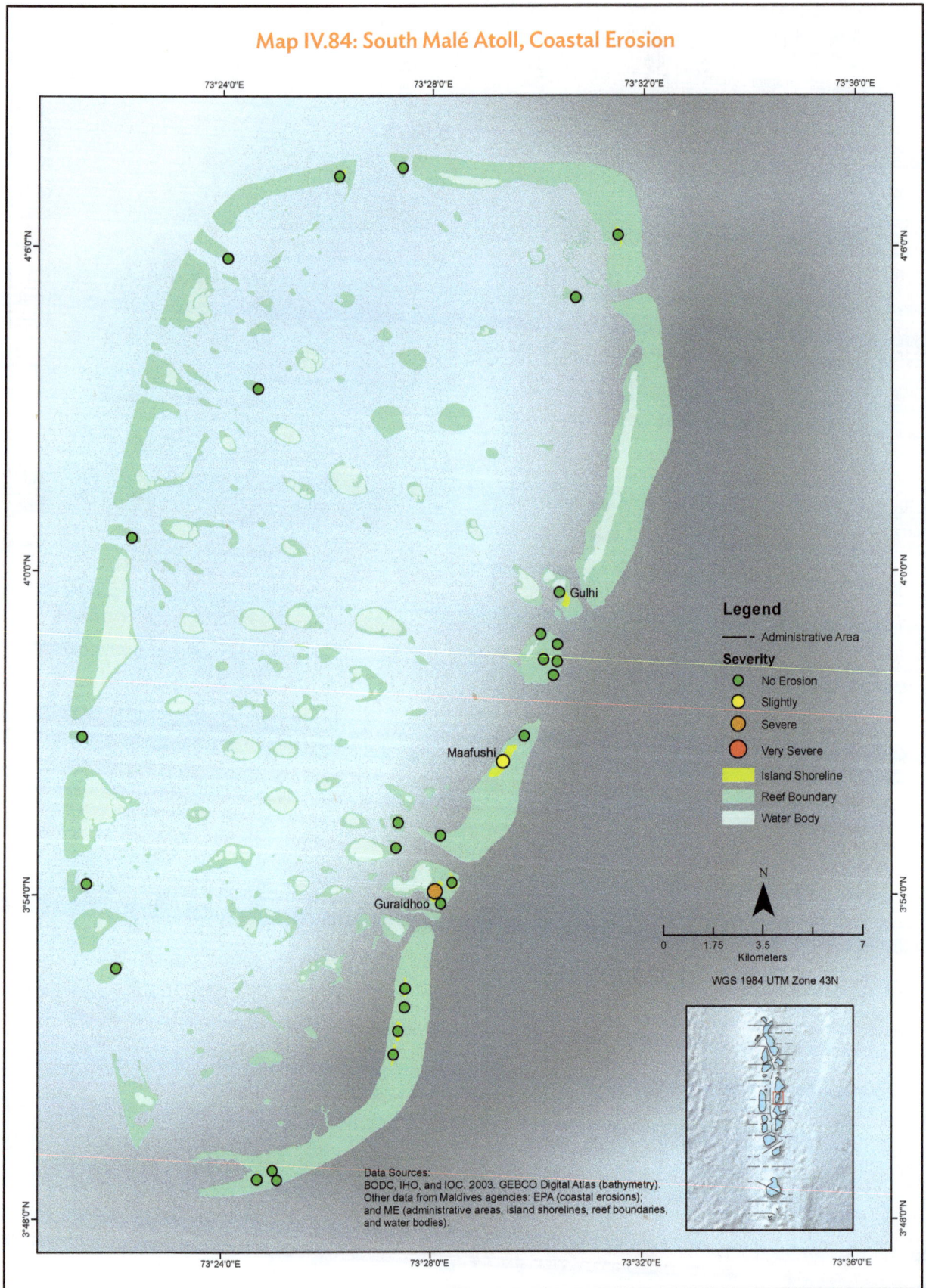

Legend

- - - Administrative Area

Severity
- ○ No Erosion
- ○ Slightly
- ○ Severe
- ○ Very Severe
- ▬ Island Shoreline
- ▬ Reef Boundary
- ▬ Water Body

N

0   1.75   3.5   7
Kilometers

WGS 1984 UTM Zone 43N

Gulhi

Maafushi

Guraidhoo

Data Sources:
BODC, IHO, and IOC. 2003. GEBCO Digital Atlas (bathymetry).
Other data from Maldives agencies: EPA (coastal erosions);
and ME (administrative areas, island shorelines, reef boundaries,
and water bodies).

Map IV.85: Thaa Atoll, Coastal Erosion

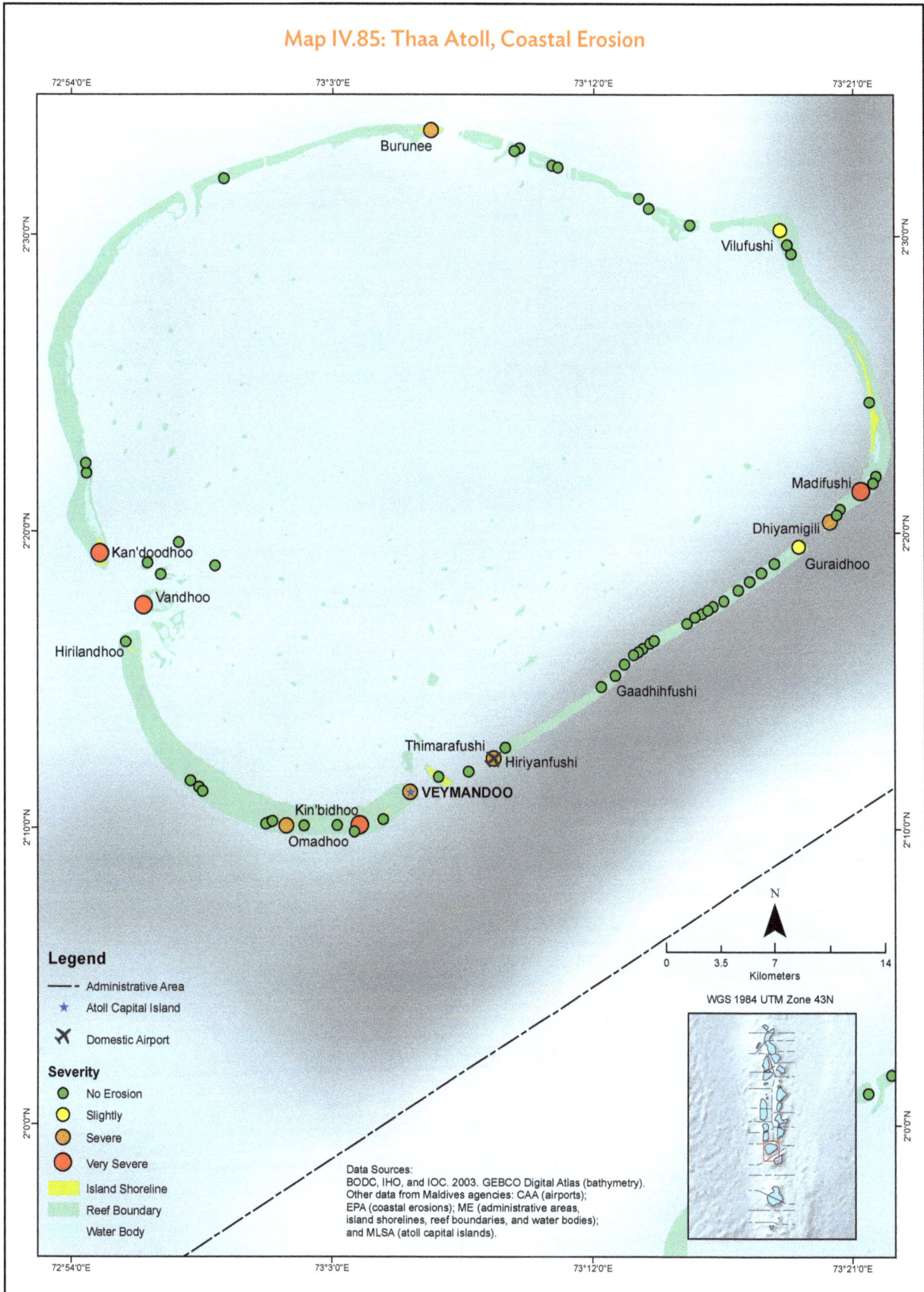

Burunee

Vilufushi

Madifushi

Dhiyamigili

Guraidhoo

Kan'doodhoo

Vandhoo

Hirilandhoo

Gaadhihfushi

Thimarafushi
Hiriyanfushi
VEYMANDOO

Kin'bidhoo
Omadhoo

N

| 0 | 3.5 | 7 | | 14 |

Kilometers

WGS 1984 UTM Zone 43N

Legend

–– Administrative Area

★ Atoll Capital Island

✈ Domestic Airport

Severity

● No Erosion

● Slightly

● Severe

● Very Severe

▮ Island Shoreline

▮ Reef Boundary

Water Body

Data Sources:
BODC, IHO, and IOC. 2003. GEBCO Digital Atlas (bathymetry).
Other data from Maldives agencies: CAA (airports);
EPA (coastal erosions); ME (administrative areas,
island shorelines, reef boundaries, and water bodies);
and MLSA (atoll capital islands).

72°54'0"E   73°3'0"E   73°12'0"E   73°21'0"E

2°30'0"N
2°20'0"N
2°10'0"N
2°0'0"N

Map IV.86: Vaavu Atoll, Coastal Erosion

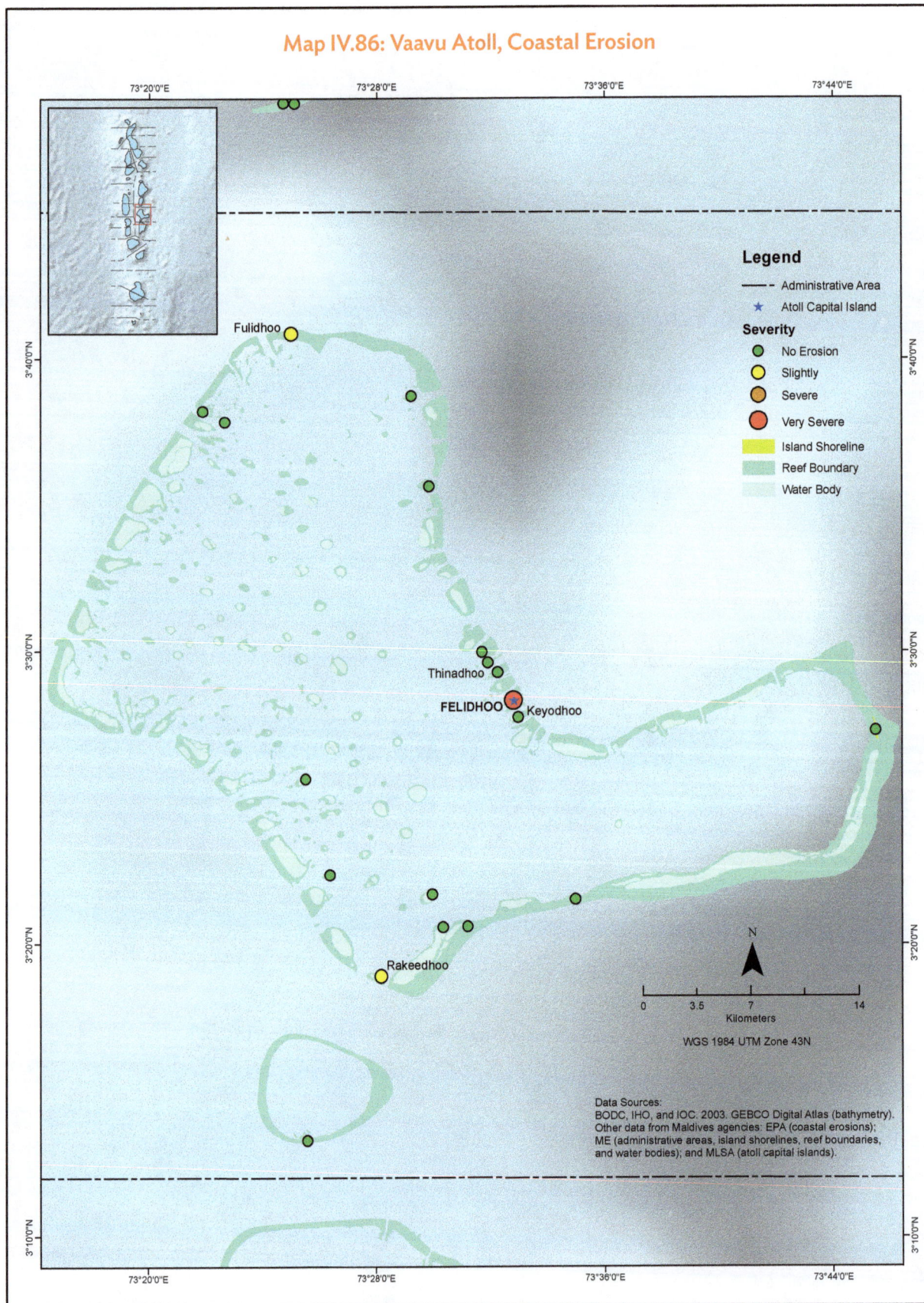

Legend

- – – Administrative Area
- ★ Atoll Capital Island

Severity
- 🟢 No Erosion
- 🟡 Slightly
- 🟠 Severe
- 🔴 Very Severe
- 🟨 Island Shoreline
- Reef Boundary
- Water Body

Fulidhoo

Thinadhoo
FELIDHOO Keyodhoo

Rakeedhoo

N

0   3.5   7   14
Kilometers

WGS 1984 UTM Zone 43N

Data Sources:
BODC, IHO, and IOC. 2003. GEBCO Digital Atlas (bathymetry).
Other data from Maldives agencies: EPA (coastal erosions);
ME (administrative areas, island shorelines, reef boundaries,
and water bodies); and MLSA (atoll capital islands).

Threats to Marine and Coastal Biodiversity

Coral Bleaching

Maldivians benefit from rich coral reefs. From serving as home to aquatic resources—the nation's main source of food—to being one of the country's tourist attractions, coral reefs are undeniably essential to life in Maldives.

Protecting corals is integral to preserving Maldives' marine resources and keeping its fishing industry alive. While corals are now better monitored, threats such as rising temperatures still cannot be controlled.

In 2015–2016, the worst coral bleaching event in Maldives happened due to high temperatures associated with the El Niño phenomenon (Ibrahim et al. 2017). Based on climate projection, annual average temperature will increase by more than 1°C in 30 years (ADB 2017). This could cause greater damage to corals in the coming years.

Coral bleaching. High seawater temperatures can cause coral bleaching, or the whitening of corals due to the loss of a symbiotic algae. Coral bleaching can lead to loss of individual corals and colonies, which in turn can cause the decline in population of numerous marine flora and fauna that depend on them.

Map IV.87: Maldives, Coral Bleaching Risk Assessment

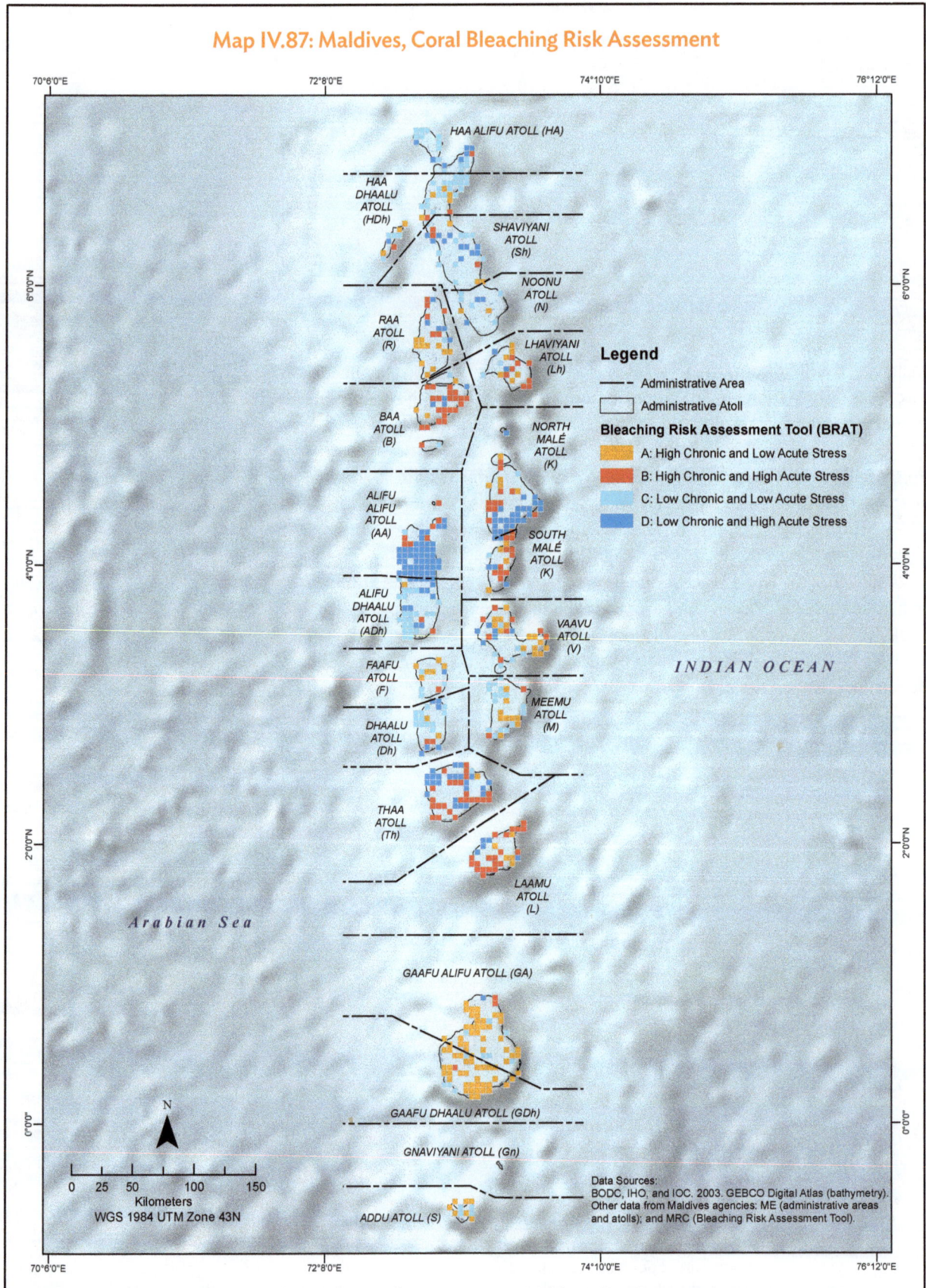

Legend

– · – Administrative Area

▢ Administrative Atoll

Bleaching Risk Assessment Tool (BRAT)

🟧 A: High Chronic and Low Acute Stress

🟥 B: High Chronic and High Acute Stress

🟦 C: Low Chronic and Low Acute Stress

🟦 D: Low Chronic and High Acute Stress

HAA ALIFU ATOLL (HA)

HAA DHAALU ATOLL (HDh)

SHAVIYANI ATOLL (Sh)

NOONU ATOLL (N)

RAA ATOLL (R)

LHAVIYANI ATOLL (Lh)

BAA ATOLL (B)

NORTH MALÉ ATOLL (K)

ALIFU ALIFU ATOLL (AA)

SOUTH MALÉ ATOLL (K)

ALIFU DHAALU ATOLL (ADh)

VAAVU ATOLL (V)

FAAFU ATOLL (F)

INDIAN OCEAN

DHAALU ATOLL (Dh)

MEEMU ATOLL (M)

THAA ATOLL (Th)

LAAMU ATOLL (L)

Arabian Sea

GAAFU ALIFU ATOLL (GA)

GAAFU DHAALU ATOLL (GDh)

GNAVIYANI ATOLL (Gn)

ADDU ATOLL (S)

N

0 25 50   100   150
Kilometers
WGS 1984 UTM Zone 43N

Data Sources:
BODC, IHO, and IOC. 2003. GEBCO Digital Atlas (bathymetry).
Other data from Maldives agencies: ME (administrative areas
and atolls); and MRC (Bleaching Risk Assessment Tool).

70°6'0"E   72°8'0"E   74°10'0"E   76°12'0"E

Disaster Risk in Maldives

The four volumes of the *Multihazard Risk Atlas of Maldives* present the various components of disaster risk in the country. *Volume I* looks at the geography of Maldives, land reclamation, and land use and land cover in the islands. *Volume II* examines the historical and projected climate that could affect Maldivians as well as their natural flora and fauna, which are then mapped out in *Volumes III and IV*.

Humans, plants, animals, and physical structures for education, health, tourism, transportation, and power are all elements exposed to natural hazards. These hazards include climate, extreme weather, earthquakes, tsunamis, typhoons, surges, sea level rise, and others. The conditions of elements such as the presence of land reclamation, sand mining activities, and coastal erosion characterize the vulnerability of the exposed islands to storm surges, sea level rise, inundation, and tsunamis. Other factors, such as the human development index, power source, health, education, and transportation, define the vulnerability of the exposed population to various hazards. Environmentally sensitive areas and bleached corals indicate an increased vulnerability of the ecosystem to environmental stresses and hazards, while having coastal protection and monitoring sites indicates adaptation capacity, which lowers vulnerability to disasters.

The following maps were prepared based on indexed risk tables, which were generated from detailed analysis of relevant data obtained from ME and the National Disaster Management Center of Maldives. These maps include physical and social risks. In addition, physical vulnerability or susceptibility maps feature multihazard hydrometeorological as well as rain-induced flooding; tsunamis; big waves or *udha*; and wave, rain, and wind hazards. Table IV.3 shows the numerical ranges of the hazard categories mapped.

Table IV.3: Hazard Categories and Index Ranges

Category	Minimum	Maximum
Rain-induced Flood		
Low	0.00	0.10
Medium	0.11	0.25
High	0.45	1.26
Udha		
Low	0.00	0.07
Medium	0.14	0.32
High	0.71	7.65
Wave, Rain, Wind (Flood) Hazard		
Low	0.00	0.05
Medium	0.10	0.33
High	0.43	1.43
Wind and Wave Hazard		
Low	0.00	0.09
Medium	0.10	0.36
High	0.45	5.26
Hydrometeorological Multihazard		
Low	0.009	
Medium	0.132	0.367
High	0.403	15.301

udha = big wave.

Source: Asian Development Bank.

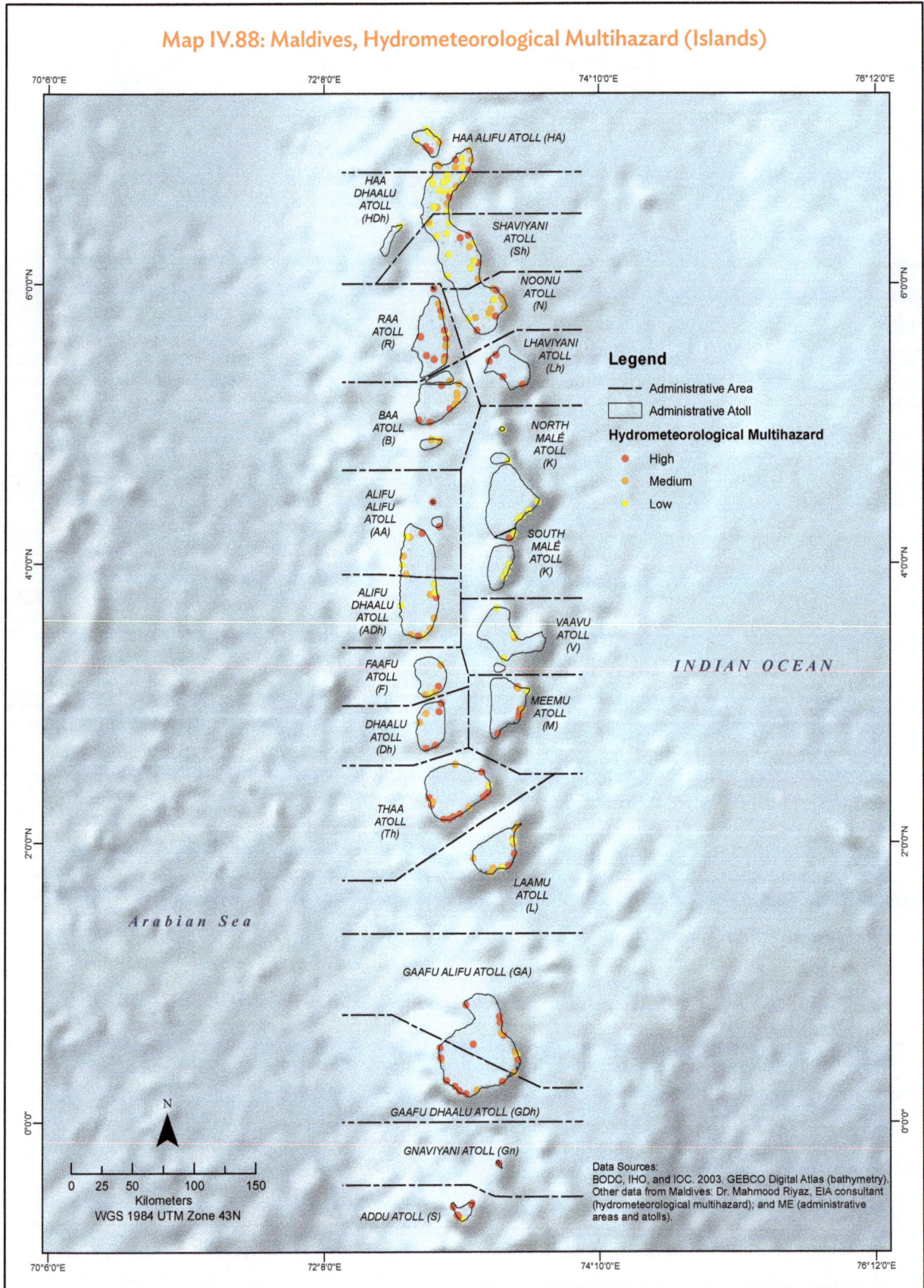

Map IV.88: Maldives, Hydrometeorological Multihazard (Islands)

HAA ALIFU ATOLL (HA)

HAA DHAALU ATOLL (HDh)

SHAVIYANI ATOLL (Sh)

NOONU ATOLL (N)

RAA ATOLL (R)

LHAVIYANI ATOLL (Lh)

BAA ATOLL (B)

NORTH MALÉ ATOLL (K)

ALIFU ALIFU ATOLL (AA)

SOUTH MALÉ ATOLL (K)

ALIFU DHAALU ATOLL (ADh)

VAAVU ATOLL (V)

FAAFU ATOLL (F)

INDIAN OCEAN

MEEMU ATOLL (M)

DHAALU ATOLL (Dh)

THAA ATOLL (Th)

LAAMU ATOLL (L)

Arabian Sea

GAAFU ALIFU ATOLL (GA)

GAAFU DHAALU ATOLL (GDh)

GNAVIYANI ATOLL (Gn)

ADDU ATOLL (S)

Legend

— · — Administrative Area

☐ Administrative Atoll

Hydrometeorological Multihazard

● High
● Medium
● Low

N

0 25 50   100   150
Kilometers
WGS 1984 UTM Zone 43N

Data Sources:
BODC, IHO, and IOC. 2003. GEBCO Digital Atlas (bathymetry).
Other data from Maldives: Dr. Mahmood Riyaz, EIA consultant
(hydrometeorological multihazard); and ME (administrative
areas and atolls).

Map IV.89: Maldives, Multihazard Physical Risk Index

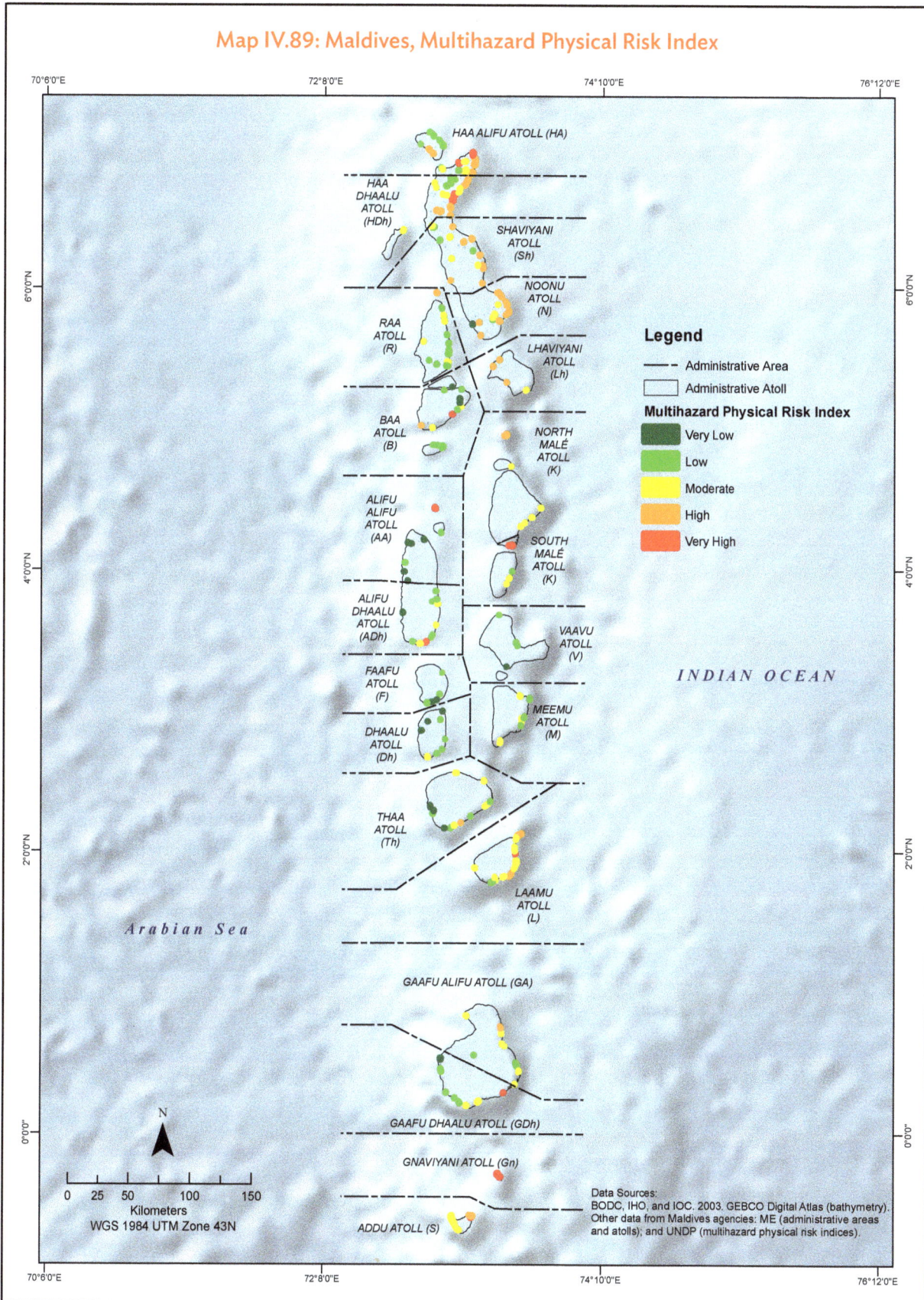

Legend

- - - Administrative Area
☐ Administrative Atoll

Multihazard Physical Risk Index
- Very Low
- Low
- Moderate
- High
- Very High

INDIAN OCEAN

Arabian Sea

N

0 25 50 100 150
Kilometers
WGS 1984 UTM Zone 43N

HAA ALIFU ATOLL (HA)
HAA DHAALU ATOLL (HDh)
SHAVIYANI ATOLL (Sh)
NOONU ATOLL (N)
RAA ATOLL (R)
LHAVIYANI ATOLL (Lh)
BAA ATOLL (B)
NORTH MALÉ ATOLL (K)
ALIFU ALIFU ATOLL (AA)
SOUTH MALÉ ATOLL (K)
ALIFU DHAALU ATOLL (ADh)
VAAVU ATOLL (V)
FAAFU ATOLL (F)
DHAALU ATOLL (Dh)
MEEMU ATOLL (M)
THAA ATOLL (Th)
LAAMU ATOLL (L)
GAAFU ALIFU ATOLL (GA)
GAAFU DHAALU ATOLL (GDh)
GNAVIYANI ATOLL (Gn)
ADDU ATOLL (S)

Data Sources:
BODC, IHO, and IOC. 2003. GEBCO Digital Atlas (bathymetry).
Other data from Maldives agencies: ME (administrative areas
and atolls); and UNDP (multihazard physical risk indices).

Map IV.90: Maldives, Multihazard Social Risk Index

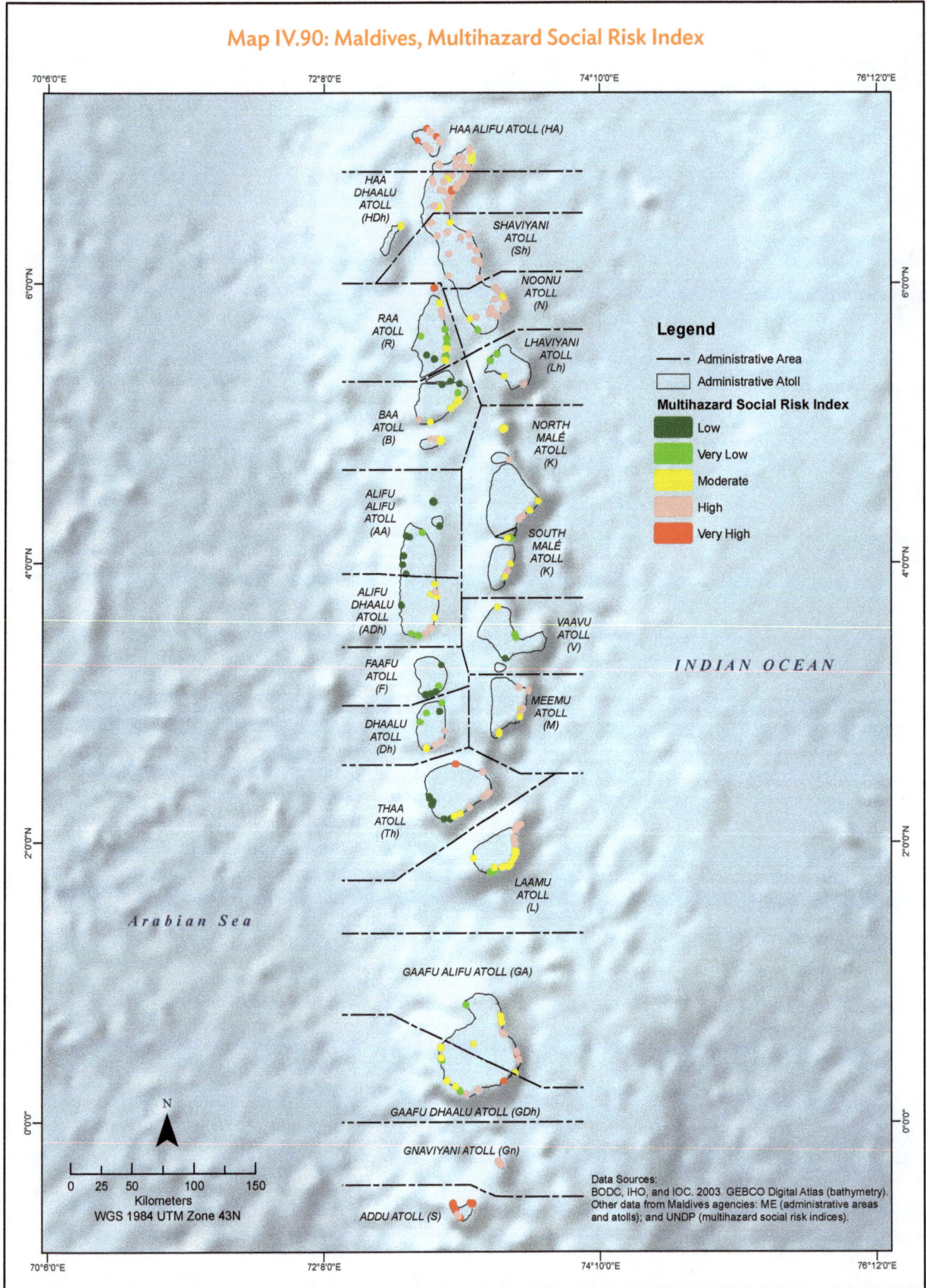

HAA ALIFU ATOLL (HA)

HAA DHAALU ATOLL (HDh)

SHAVIYANI ATOLL (Sh)

NOONU ATOLL (N)

RAA ATOLL (R)

LHAVIYANI ATOLL (Lh)

BAA ATOLL (B)

NORTH MALÉ ATOLL (K)

ALIFU ALIFU ATOLL (AA)

SOUTH MALÉ ATOLL (K)

ALIFU DHAALU ATOLL (ADh)

VAAVU ATOLL (V)

FAAFU ATOLL (F)

MEEMU ATOLL (M)

DHAALU ATOLL (Dh)

THAA ATOLL (Th)

LAAMU ATOLL (L)

GAAFU ALIFU ATOLL (GA)

GAAFU DHAALU ATOLL (GDh)

GNAVIYANI ATOLL (Gn)

ADDU ATOLL (S)

INDIAN OCEAN

Arabian Sea

Legend

— · — Administrative Area

▢ Administrative Atoll

Multihazard Social Risk Index
- Low
- Very Low
- Moderate
- High
- Very High

Kilometers 0 25 50 100 150
WGS 1984 UTM Zone 43N

Data Sources:
BODC, IHO, and IOC. 2003. GEBCO Digital Atlas (bathymetry). Other data from Maldives agencies: ME (administrative areas and atolls); and UNDP (multihazard social risk indices).

Map IV.91: Maldives, Rain-Induced Flood (Islands)

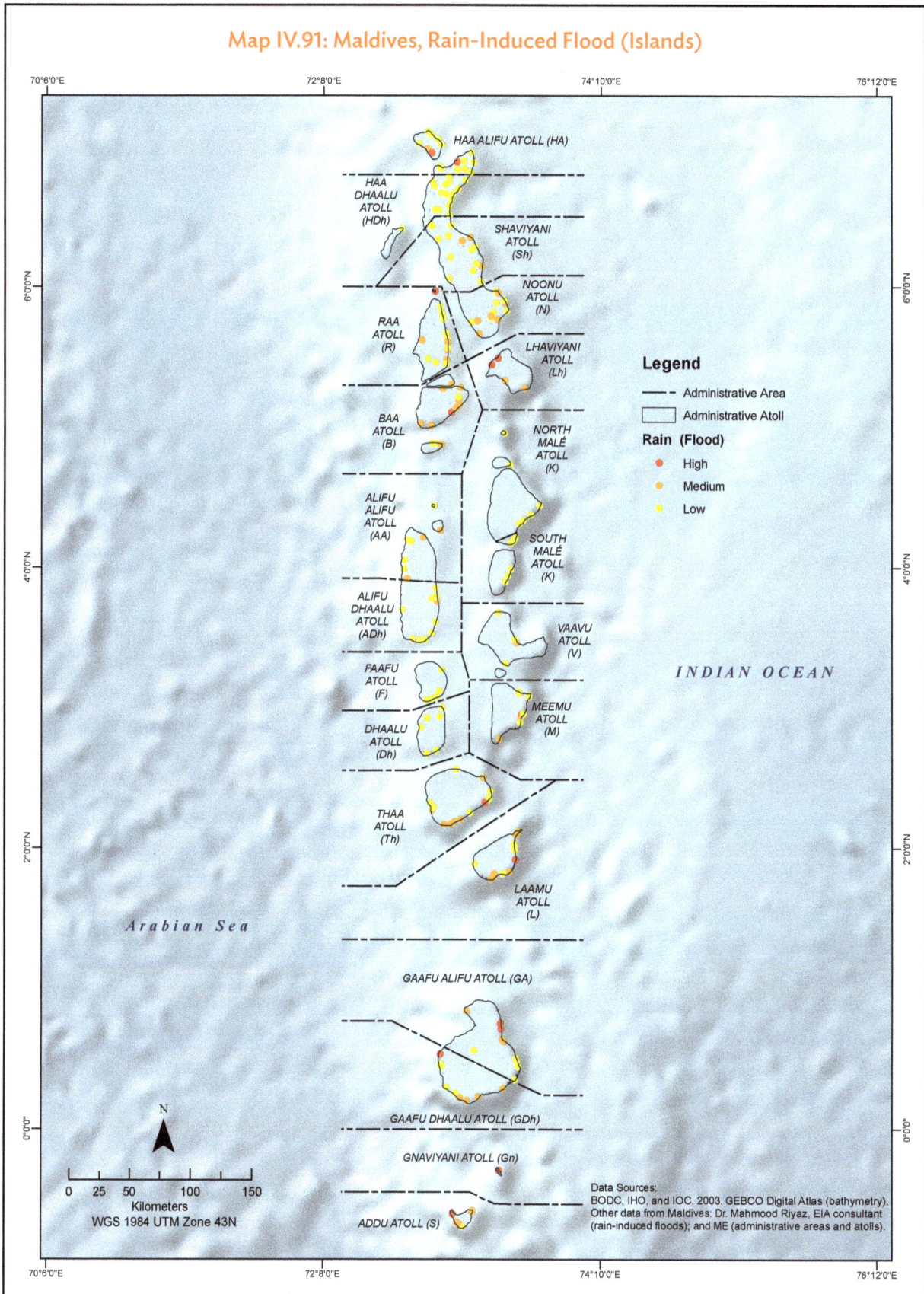

Legend

- - - Administrative Area
☐ Administrative Atoll

Rain (Flood)
- 🔴 High
- 🟠 Medium
- 🟡 Low

HAA ALIFU ATOLL (HA)
HAA DHAALU ATOLL (HDh)
SHAVIYANI ATOLL (Sh)
NOONU ATOLL (N)
RAA ATOLL (R)
LHAVIYANI ATOLL (Lh)
BAA ATOLL (B)
NORTH MALÉ ATOLL (K)
ALIFU ALIFU ATOLL (AA)
SOUTH MALÉ ATOLL (K)
ALIFU DHAALU ATOLL (ADh)
VAAVU ATOLL (V)
FAAFU ATOLL (F)
MEEMU ATOLL (M)
DHAALU ATOLL (Dh)
THAA ATOLL (Th)
LAAMU ATOLL (L)
GAAFU ALIFU ATOLL (GA)
GAAFU DHAALU ATOLL (GDh)
GNAVIYANI ATOLL (Gn)
ADDU ATOLL (S)

INDIAN OCEAN
Arabian Sea

N

0 25 50 100 150
Kilometers
WGS 1984 UTM Zone 43N

Data Sources:
BODC, IHO, and IOC. 2003. GEBCO Digital Atlas (bathymetry). Other data from Maldives: Dr. Mahmood Riyaz, EIA consultant (rain-induced floods); and ME (administrative areas and atolls).

Map IV.92: Maldives, Tsunami Hazard Rank (Islands)

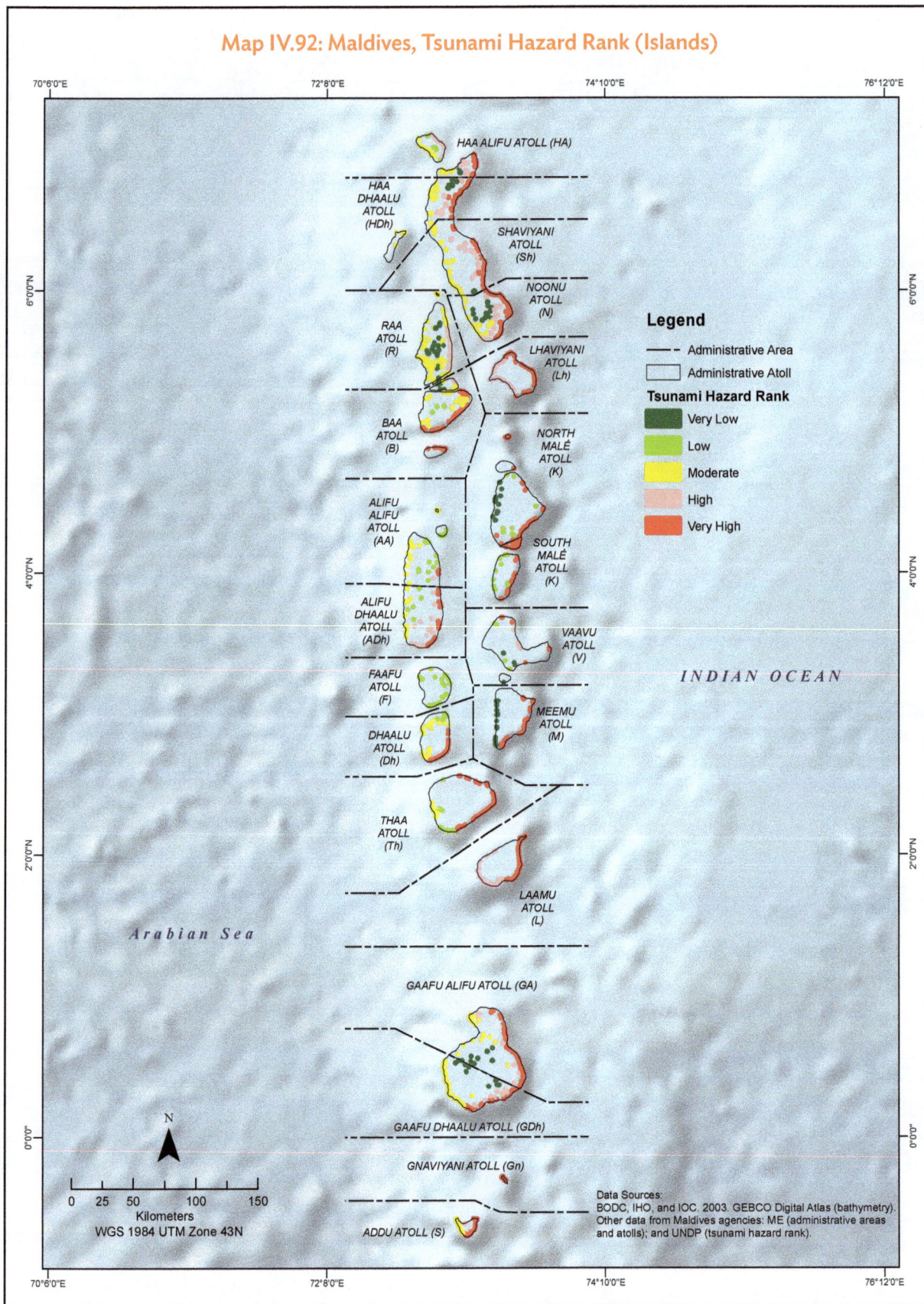

Legend

- – – Administrative Area
- ☐ Administrative Atoll

Tsunami Hazard Rank
- Very Low
- Low
- Moderate
- High
- Very High

HAA ALIFU ATOLL (HA)
HAA DHAALU ATOLL (HDh)
SHAVIYANI ATOLL (Sh)
NOONU ATOLL (N)
RAA ATOLL (R)
LHAVIYANI ATOLL (Lh)
BAA ATOLL (B)
NORTH MALÉ ATOLL (K)
ALIFU ALIFU ATOLL (AA)
SOUTH MALÉ ATOLL (K)
ALIFU DHAALU ATOLL (ADh)
VAAVU ATOLL (V)
FAAFU ATOLL (F)
MEEMU ATOLL (M)
DHAALU ATOLL (Dh)
THAA ATOLL (Th)
LAAMU ATOLL (L)
GAAFU ALIFU ATOLL (GA)
GAAFU DHAALU ATOLL (GDh)
GNAVIYANI ATOLL (Gn)
ADDU ATOLL (S)

INDIAN OCEAN
Arabian Sea

N

0 25 50 100 150
Kilometers
WGS 1984 UTM Zone 43N

Data Sources:
BODC, IHO, and IOC. 2003. GEBCO Digital Atlas (bathymetry).
Other data from Maldives agencies: ME (administrative areas and atolls); and UNDP (tsunami hazard rank).

Map IV.93: Maldives, *Udha* Hazard (Islands)

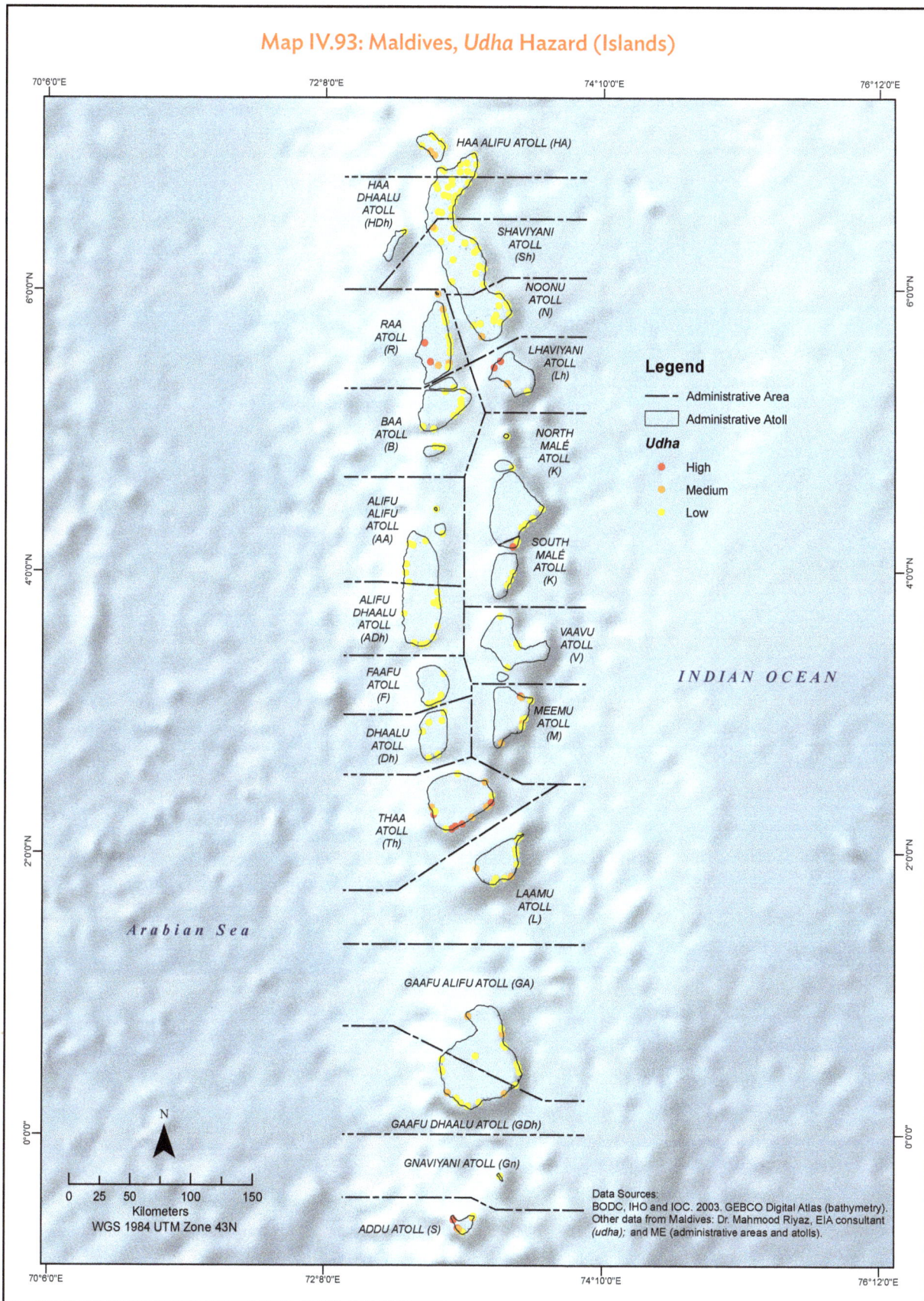

70°6'0"E   72°8'0"E   74°10'0"E   76°12'0"E

HAA ALIFU ATOLL (HA)

HAA DHAALU ATOLL (HDh)

SHAVIYANI ATOLL (Sh)

NOONU ATOLL (N)

RAA ATOLL (R)

LHAVIYANI ATOLL (Lh)

BAA ATOLL (B)

NORTH MALÉ ATOLL (K)

ALIFU ALIFU ATOLL (AA)

SOUTH MALÉ ATOLL (K)

ALIFU DHAALU ATOLL (ADh)

VAAVU ATOLL (V)

FAAFU ATOLL (F)

INDIAN OCEAN

DHAALU ATOLL (Dh)

MEEMU ATOLL (M)

THAA ATOLL (Th)

LAAMU ATOLL (L)

Arabian Sea

GAAFU ALIFU ATOLL (GA)

GAAFU DHAALU ATOLL (GDh)

GNAVIYANI ATOLL (Gn)

ADDU ATOLL (S)

Legend

— · — Administrative Area

▭ Administrative Atoll

Udha

● High

● Medium

● Low

N

0   25   50   100   150
Kilometers
WGS 1984 UTM Zone 43N

Data Sources:
BODC, IHO and IOC. 2003. GEBCO Digital Atlas (bathymetry).
Other data from Maldives: Dr. Mahmood Riyaz, EIA consultant
(*udha*); and ME (administrative areas and atolls).

Map IV.94: Maldives, Wind and Wave Hazards

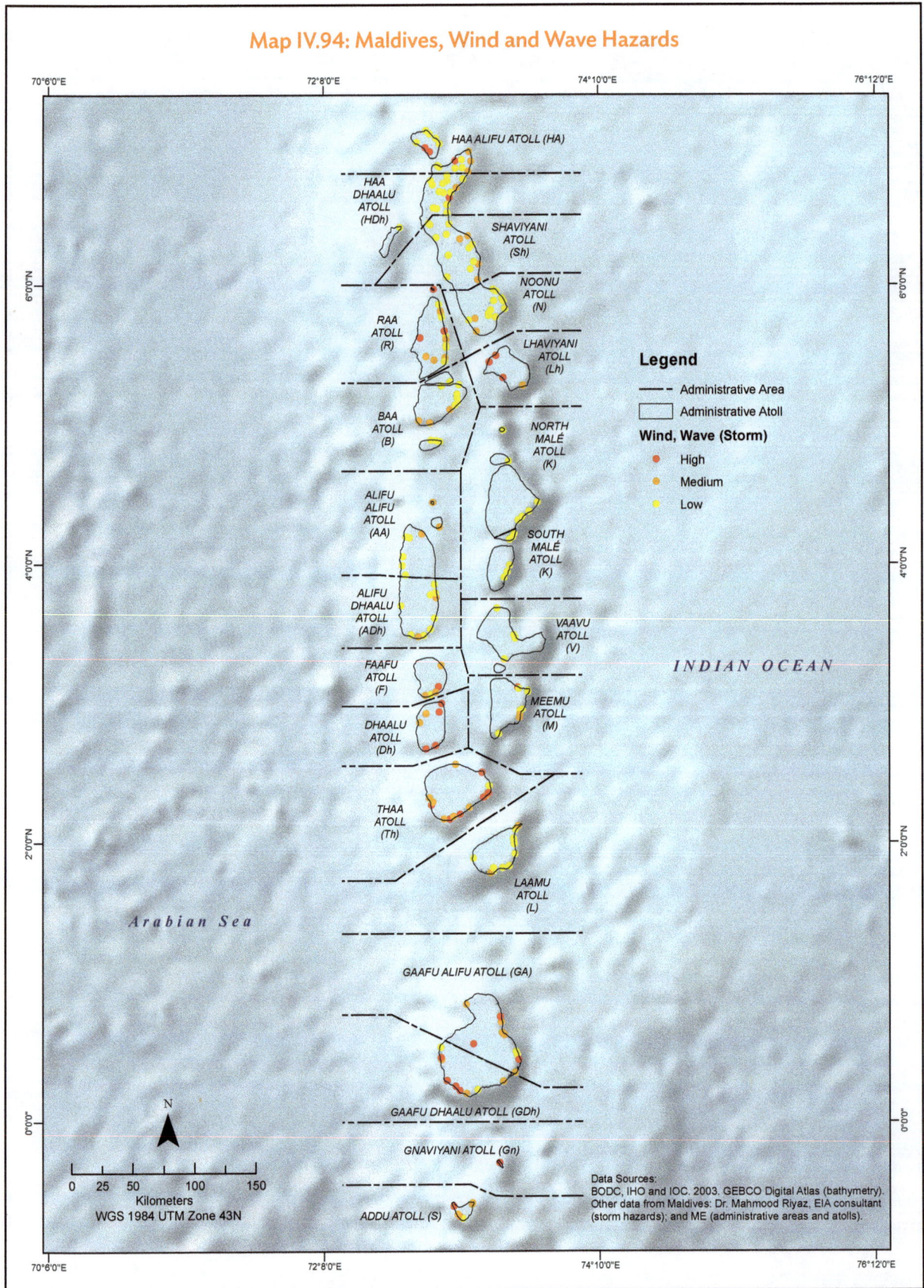

Legend

- ⎯ ⋅ ⎯ Administrative Area
- ▭ Administrative Atoll

Wind, Wave (Storm)
- 🔴 High
- 🟠 Medium
- 🟡 Low

HAA ALIFU ATOLL (HA)
HAA DHAALU ATOLL (HDh)
SHAVIYANI ATOLL (Sh)
NOONU ATOLL (N)
RAA ATOLL (R)
LHAVIYANI ATOLL (Lh)
BAA ATOLL (B)
NORTH MALÉ ATOLL (K)
ALIFU ALIFU ATOLL (AA)
SOUTH MALÉ ATOLL (K)
ALIFU DHAALU ATOLL (ADh)
VAAVU ATOLL (V)
FAAFU ATOLL (F)
MEEMU ATOLL (M)
DHAALU ATOLL (Dh)
THAA ATOLL (Th)
LAAMU ATOLL (L)
GAAFU ALIFU ATOLL (GA)
GAAFU DHAALU ATOLL (GDh)
GNAVIYANI ATOLL (Gn)
ADDU ATOLL (S)

INDIAN OCEAN
Arabian Sea

Kilometers 0 25 50 100 150
WGS 1984 UTM Zone 43N

Data Sources:
BODC, IHO and IOC. 2003. GEBCO Digital Atlas (bathymetry). Other data from Maldives: Dr. Mahmood Riyaz, EIA consultant (storm hazards); and ME (administrative areas and atolls).

Map Data Sources

Government Ministries, Departments, and Agencies in Maldives
    Civil Aviation Authority
        Airports
    Environmental Protection Agency
        Coastal erosion
        Environmentally protected and sensitive areas
    Land and Survey Authority
        Atoll capital islands
        Cities
    Marine Research Centre
        Coral reef monitoring sites
    Ministry of Economic Development
        Ports
    Ministry of Environment
        Administrative areas
        Administrative atolls
        Island shorelines
        Mangroves
        Reef boundaries
        Water bodies
    Ministry of Fisheries, Marine Resources and Agriculture
        Fish aggregating devices
    Ministry of National Planning and Infrastructure
        Coastal protection

International Institutions
    Marine Spatial Ecology Lab, University of Queensland, Australia
        Bleaching Risk Assessment Tool

Private Individual
    Mahmood Riyaz, EPA-Licensed Environmental Impact Assessment Specialist
        Hydrometeorological multihazard
        Rain-induced flood
        Udha vulnerability
        Wave, rain, wind (flood) vulnerability
        Wind, wave (storm) vulnerability

References

Asian Development Bank. 2017. *Final Report on Climate Downscaling for Bangladesh, Maldives, and Sri Lanka.* Consultants' report. Manila (TA 8572-REG).

British Oceanographic Data Centre (BODC), International Hydrographic Organisation (IHO) and the Intergovernmental Oceanographic Commission (IOC) of the United Nations Educational, Scientific and Cultural Organization. 2003. *General Bathymetric Chart of the Oceans (GEBCO) Digital Atlas.* UK: British Oceanographic Data Centre.

Emerton, L., S. Baig, and M. Saleem. 2009. *Valuing Biodiversity: The Economic Case for Biodiversity Conservation in the Maldives.* Homagama: Ecosystems and Livelihoods Group Asia; International Union for the Conservation of Nature for the Atoll Ecosystem Conservation Project; Ministry of Housing, Transport, and Environment, Government of Maldives.

Ibrahim, N., M. Mohamed, A. Basheer, H. Ismail, F. Nistharan, A. Schmidt, R. Naeem, A. Abdulla, and G. Grimsditch. 2017. *Status of Coral Bleaching in the Maldives in 2016.* Malé: Marine Research Centre.

United Nations Development Programme. 2006. *Developing a Disaster Risk Profile for Maldives.*

Maps were prepared by the Country Consultant Team and the Manila Observatory on behalf of the Asian Development Bank.